Multi-Agent Machine Learning

Multi-Agent Machine Learning
A Reinforcement Approach

Howard M. Schwartz
Department of Systems and Computer Engineering
Carleton University

Published by John Wiley & Sons, Inc., Hoboken, New Jersey
Published simultaneously in Canada

For general information on our other products and services or for technical support, please contact our Customer Care Department within the United States at (800) 762-2974, outside the United States at (317) 572-3993 or fax (317) 572-4002.

Wiley also publishes its books in a variety of electronic formats. Some content that appears in print may not be available in electronic formats. For more information about Wiley products, visit our web site at www.wiley.com.

Library of Congress Cataloging-in-Publication Data:

Schwartz, Howard M., editor.
  Multi-agent machine learning : a reinforcement approach / Howard M. Schwartz.
    pages cm
  Includes bibliographical references and index.
  ISBN 978-1-118-36208-2 (hardback)
  1. Reinforcement learning. 2. Differential games. 3. Swarm intelligence. 4. Machine learning. I. Title.
  Q325.6.S39 2014
  519.3–dc23
                                                                                2014016950

10 9 8 7 6 5 4 3 2 1

Contents

**Chapter 6 Swarm Intelligence and the Evolution
of Personality Traits** **200**

Preface

For a decade I have taught a course on adaptive control. The course focused on the classical methods of system identification, using such classic texts as Ljung [1, 2]. The course addressed traditional methods of model reference adaptive control and nonlinear adaptive control using Lyapunov techniques. However, the theory had become out of sync with current engineering practice. As such, my own research and the focus of the graduate course changed to include adaptive signal processing, and to incorporate adaptive channel equalization and echo cancellation using the least mean squares (LMS) algorithm. The course name likewise changed, from "Adaptive Control" to "Adaptive and Learning Systems." My research was still focused on system identification and nonlinear adaptive control with application to robotics. However, by the early 2000s, I had started work with teams of robots. It was now possible to use handy robot kits and low-cost microcontroller boards to build several robots that could work together. The graduate course in adaptive and learning systems changed again; the theoretical material on nonlinear adaptive control using Lyapunov techniques was reduced, replaced with ideas from reinforcement learning. A whole new range of applications developed. The teams of robots had to learn to work together and to compete.

Today, the graduate course focuses on system identification using recursive least squares techniques, some model reference adaptive control (still using Lyapunov techniques), adaptive signal processing using the LMS algorithm, and reinforcement learning using Q-learning. The first two chapters of this book present these ideas in an abridged form, but in sufficient detail to demonstrate the connections among the learning algorithms that are available; how they are the same; and how they are different. There are other texts that cover this material in detail [2–4].

The research then began to focus on teams of robots learning to work together. The work examined applications of robots working together for search and rescue applications, securing important infrastructure and border regions. It also began to focus on reinforcement learning and multiagent reinforcement learning. The robots are the learning agents. How do children learn how to play tag? How do we learn to play football, or how do police work together to capture a criminal? What strategies do we use, and how do we formulate these strategies? Why can I play touch football with a new group of people and quickly be able to assess everyone's capabilities and then take a particular strategy in the game?

As our research team began to delve further into the ideas associated with multiagent machine learning and game theory, we discovered that the published literature covered many ideas but was poorly coordinated or focused. Although there are a few survey articles [5], they do not give sufficient details to appreciate the different methods. The purpose of this book is to introduce the reader to a particular form of machine learning. The book focuses on multiagent machine learning, but it is tied together with the central theme of learning algorithms in general. Learning algorithms come in many different forms. However, they tend to have a similar approach. We will present the differences and similarities of these methods.

This book is based on my own work and the work of several doctoral and masters students who have worked under my supervision over the past 10 years. In particular, I would like to thank Prof. Sidney Givigi. Prof. Givigi was instrumental in developing the ideas and algorithms presented in Chapter 6. The doctoral research of Xiaosong (Eric) Lu has also found its way into this book. The work on guarding a territory is largely based on his doctoral dissertation. Other graduate students who helped me in this work include Badr Al Faiya, Mostafa Awheda, Pascal De Beck-Courcelle, and Sameh Desouky. Without the dedicated work of this group of students, this book would not have been possible.

H. M. Schwartz
Ottawa, Canada
September, 2013

References

[1] L. Ljung, *System Identification: Theory for the User*. Upper Saddle River, NJ: Prentice Hall, 2nd ed., 1999.

[2] L. Ljung and T. Soderstrom, *Theory and Practice of Recursive Identification.* Cambridge, Massachusetts: The MIT Press, 1983.

[3] R. S. Sutton and A. G. Barto, *Reinforcement learning: An Introduction.* Cambridge, Massachusetts: The MIT Press, 1998.

[4] Astrom, K. J. and Wittenmark, B., *Adaptive Control.* Boston, Massachusetts: Addison-Wesley Longman Publishing Co., Inc., 2nd ed., 1994, ISBN = 0201558661.

[5] L. Buşoniu and R. Babuška, and B. D. Schutter, "A comprehensive survey of multiagent reinforcement learning," *IEEE Trans. Syst. Man Cybern. Part C*, Vol. 38, no. 2, pp. 156–172, 2008.

Chapter 1
A Brief Review of Supervised Learning

There are a number of algorithms that are typically used for system identification, adaptive control, adaptive signal processing, and machine learning.
These algorithms all have particular similarities and differences. However, they
all need to process some type of experimental data. How we collect the data
and process it determines the most suitable algorithm to use. In adaptive control, there is a device referred to as the *self-tuning regulator*. In this case, the
algorithm measures the states as outputs, estimates the model parameters, and
outputs the control signals. In reinforcement learning, the algorithms process
rewards, estimate value functions, and output actions. Although one may refer
to the recursive least squares (RLS) algorithm in the self-tuning regulator as a
supervised learning algorithm and reinforcement learning as an unsupervised
learning algorithm, they are both very similar. In this chapter, we will present
a number of well-known baseline supervised learning algorithms.

1.1 Least Squares Estimates

The least squares (LS) algorithm is a well-known and robust algorithm for
fitting experimental data to a model. The first step is for the user to define
a mathematical structure or model that he/she believes will fit the data. The
second step is to design an experiment to collect data under suitable conditions.
"Suitable conditions" usually means the operating conditions under which the

Multi-Agent Machine Learning: A Reinforcement Approach, First Edition. Howard M. Schwartz.
© 2014 John Wiley & Sons, Inc. Published 2014 by John Wiley & Sons, Inc.

system will typically operate. The next step is to run the estimation algorithm, which can take several forms, and, finally, validate the identified or "learned" model. The LS algorithm is often used to fit the data. Let us look at the case of the classical two-dimensional linear regression fit that we are all familiar with:

$$y(n) = ax(n) + b \tag{1.1}$$

In this a simple linear regression model, where the input is the sampled signal $x(n)$ and the output is $y(n)$. The model structure defined is a straight line. Therefore, we are assuming that the data collected will fit a straight line. This can be written in the form

$$y(n) = \phi^T \theta \tag{1.2}$$

where $\phi^T = \begin{bmatrix} x(n) & 1 \end{bmatrix}$ and $\theta^T = \begin{bmatrix} a & b \end{bmatrix}$. How one chooses ϕ determines the model structure, and this reflects how one believes the data should behave. This is the essence of machine learning, and virtually all university students will at some point learn the basic statistics of linear regression. Behind the computations of the linear regression algorithm is the scalar cost function, given by

$$V = \sum_{n=1}^{N} (y(n) - \phi^T(n)\hat{\theta})^2 \tag{1.3}$$

The term $\hat{\theta}$ is the estimate of the LS parameter θ. The goal is for the estimate $\hat{\theta}$ to minimize the cost function V. To find the "optimal" value of the parameter estimate $\hat{\theta}$, one takes the partial derivative of the cost function V with respect to $\hat{\theta}$ and sets this derivative to zero. Therefore, one gets

$$\frac{\partial V}{\partial \hat{\theta}} = \sum_{n=1}^{N} (y(n) - \phi^T(n)\hat{\theta})\phi(n)$$

$$= \sum_{n=1}^{N} \phi(n)y(n) - \sum_{n=1}^{N} \phi(n)\phi^T(n)\hat{\theta} \tag{1.4}$$

Setting $\dfrac{\partial V}{\partial \hat{\theta}} = 0$, we get

$$\sum_{n=1}^{N} \phi(n)\phi^T(n)\hat{\theta} = \sum_{n=1}^{N} \phi(n)y(n) \tag{1.5}$$

Solving for $\hat{\theta}$, we get the LS solution

$$\hat{\theta} = \left[\sum_{n=1}^{N} \phi(n)\phi^T(n) \right]^{-1} \left[\sum_{n=1}^{N} \phi(n)y(n) \right] \tag{1.6}$$

where the inverse, $\left[\sum_{n=1}^{N} \phi(n)\phi^T(n) \right]^{-1}$, exists. If the inverse does not exist, then the system is not identifiable. For example, if in the straight line case one only had a single point, then the inverse would not span the two-dimensional space and it would not exist. One needs at least two independent points to draw a straight line. Or, for example, if one had exactly the same point over and over again, then the inverse would not exist. One needs at least two independent points to draw a straight line. The matrix $\left[\sum_{n=1}^{N} \phi(n)\phi^T(n) \right]$ is referred to as the *information matrix* and is related to how well one can estimate the parameters. The inverse of the information matrix is the covariance matrix, and it is proportional to the variance of the parameter estimates. Both these matrices are positive definite and symmetric. These are very important properties which are used extensively in analyzing the behavior of the algorithm. In the literature, one will often see the covariance matrix referred to as $P = \left[\sum_{n=1}^{N} \phi(n)\phi^T(n) \right]^{-1}$. We can write the second equation on the right of Eq. (1.4) in the form

$$\frac{\partial V}{\partial \hat{\theta}} = 0 = \sum_{n=1}^{N} (y(n) - \phi^T(n)\hat{\theta})\phi(n) \tag{1.7}$$

and one can define the prediction errors as

$$\epsilon(n) = (y(n) - \phi^T(n)\hat{\theta}) \tag{1.8}$$

The term within brackets in Eq. (1.7) is known as the *prediction error* or, as some people will refer to it, the *innovations*. The term $\epsilon(n)$ represents the error in predicting the output of the system. In this case, the output term $y(n)$ is the correct answer, which is what we want to estimate. Since we know the correct answer, this is referred to as *supervised learning*. Notice that the value of the prediction error times the data vector is equal to zero. We then say that the prediction errors are orthogonal to the data, or that the data sits in the null space of the prediction errors. In simplistic terms, this means that, if one has chosen a good model structure $\phi(n)$, then the prediction errors should appear as white noise. Always plot the prediction errors as a quick check to see how good your predictor is. If the errors appear to be correlated (i.e., not white noise), then you can improve your model and get a better prediction.

One does not typically write the linear regression in the form of Eq. (1.2), but typically will add a white noise term, and then the linear regression takes the form

$$y(n) = \phi^T(n)\theta + v(n) \tag{1.9}$$

where $v(n)$ is a white noise term. Equation (1.9) can represent an infinite number of possible model structures. For example, let us assume that we want to learn the dynamics of a second-order linear system or the parameters of a second-order infinite impulse response (IIR) filter. Then we could choose the second-order model structure given by

$$y(n) = -a_1 y(n-1) - a_2 y(n-2) + b_1 u(n-1) + b_2 u(n-2) + v(n) \tag{1.10}$$

Then the model structure would be defined in $\phi(n)$ as

$$\phi^T(n) = \begin{bmatrix} y(n-1) & y(n-2) & u(n-1) & u(n-2) \end{bmatrix} \tag{1.11}$$

In general, one can write an arbitrary kth-order autoregressive exogenous (ARX) model structure as

$$y(n) = -a_1 y(n-1) - a_2 y(n-2) - \cdots - a_m y(n-k)$$
$$+ b_1 u(n-1) + b_2 u(n-2) + \cdots + b_{n-k} u(n-k) + v(n) \tag{1.12}$$

and $\phi(n)$ takes the form

$$\phi^T(n) = \begin{bmatrix} y(n-1) & \cdots & y(n-m) & u(n-1) & \cdots & u(n-m) \end{bmatrix} \tag{1.13}$$

One then collects the data from a suitable experiment (easier said than done!), and then computes the parameters using Eq. (1.6). The vector $\phi(n)$ can take many different forms; in fact, it can contain nonlinear functions of the data, for example, logarithmic terms or square terms, and it can have different delay terms. To a large degree, one can use ones professional judgment as to what to put into $\phi(n)$. One will often write the data in the matrix form, in which case the matrix is defined as

$$\Phi = \begin{bmatrix} \phi(1) & \phi(2) & \cdots & \phi(N) \end{bmatrix} \tag{1.14}$$

and the output matrix as

$$Y = \begin{bmatrix} y(1) & y(2) & \cdots & y(N) \end{bmatrix} \tag{1.15}$$

Then one can write the LS estimate as

$$\hat{\Theta} = (\Phi\Phi^T)^{-1}\Phi Y \tag{1.16}$$

Furthermore, one can write the prediction errors as

$$E = Y - \Phi^T\hat{\Theta} \tag{1.17}$$

We can also write the orthogonality condition as $\Phi E = 0$.

The LS method of parameter identification or machine learning is very well developed and there are many properties associated with the technique. In fact, much of the work in statistical inference is derived from the few equations described in this section. This is the beginning of many scientific investigations including work in the social sciences.

1.2 Recursive Least Squares

The LS algorithm has been extended to the RLS algorithm. In this case, the parameter estimate is developed as the machine collects the data in real time. In the previous section, all the data was collected first, and then the parameter estimates were computed on the basis of Eq. (1.6). The RLS algorithm is derived by assuming a solution to the LS algorithm and then adding a single data point. The derivation is shown in Reference 1. In the RLS implementation, the cost function takes a slightly different form. The cost function in this case is

$$V = \sum_{n=1}^{N} \lambda^{(N-t)}(y(n) - \phi^T(n)\hat{\theta})^2 \tag{1.18}$$

where $\lambda \leq 1$. The term λ is known as the *forgetting factor*. This term will place less weight on older data points. As such, the resulting RLS algorithm will be able track changes to the parameters. Once again, taking the partial derivative of V with respect to $\hat{\theta}$ and setting the derivative to zero, we get

$$\hat{\theta} = \left[\sum_{n=1}^{N} \lambda^{(N-t)}\phi(n)\phi^T(n)\right]^{-1}\left[\sum_{n=1}^{N} \lambda^{(N-t)}\phi(n)y(n)\right] \tag{1.19}$$

The forgetting factor should be set as $0.95 \leq \lambda \leq 1.0$. If one sets the forgetting factor near 0.95, then old data is forgotten very quickly; the rule of thumb is that the estimate of the parameters $\hat{\theta}$ is approximately based on $1/(1 - \lambda)$ data points.

The RLS algorithm is as follows:

$$\hat{\theta}(n+1) = \hat{\theta}(n) + L(n+1)(y(n+1) - \phi^T(n+1)\hat{\theta}(n))$$

$$L(n+1) = \frac{P(n)\phi(n+1)}{\lambda + \phi^T(n+1)P(n)\phi(n+1)}$$

$$P(n+1) = \frac{1}{\lambda}\left(P(n) - \frac{P(n)\phi(n+1)\phi^T(n+1)P(n)}{\lambda + \phi^T(n+1)P(n)\phi(n+1)}\right) \qquad (1.20)$$

One implements Eq. (1.20) by initializing the parameter estimation vector $\hat{\theta}$ to the users best initial estimate of the parameters, which is often simply zero. The covariance matrix P is typically initialized to a relatively large diagonal matrix, and represents the initial uncertainty in the parameter estimate.

One can implement the RLS algorithm as in Eq. (1.20), but the user should be careful that the covariance matrix P is always positive definite and symmetric. If the P matrix, because of numerical error by repeatedly computing the RLS, ceases to be positive definite and symmetric, then the algorithm will diverge. There are a number of well-developed algorithms to ensure that the P matrix remains positive definite. One can use a square-roots approach whereby the P matrix is factored into its Cholesky factorization or the UDU factorization. Such methods are described in Reference 1.

Let us examine Eq. (1.20) and notice that the update to the parameter estimate is the previous estimate plus a matrix $L(n)$ times the current prediction error. We will see this structure in almost every algorithm that will be described in machine learning. In this case, we have an actual correct answer, which is the measurement $y(n)$, and we call such algorithms *supervised learning*.

1.3 Least Mean Squares

In the field of signal processing, there are a few commonly used techniques to model or characterize the dynamics of a communications channel and then compensate for the effects of the channel on the signal. These techniques are referred to as *channel equalization* and *echo cancellation*. There are numerous books on adaptive signal processing and adaptive filtering [2]. Most of these techniques use the least mean squares (LMS) approach to identify the coefficients of a model of the channel. Once again, as in the LS and RLS algorithms, we must choose an appropriate model structure to define the communication channel dynamics. In the field of signal processing, one would typically use what is known as a *finite impulse response* (FIR) filter as the underlying model structure that describes the system. To maintain consistency with the previous section, one can write the channel dynamics as

$$y(n) = b_0 u(n) + b_1 u(n-1) + \cdots + b_k u(n-k) + v(n) \qquad (1.21)$$

where $y(n)$ is the output of the filter, or the communications channel, at time step n, b_i are the filter coefficients that we want to estimate or *learn*, and $u(n)$ is the input signal. Typically, the signal $u(n)$ is the communication signal that we want to recover from the output signal $y(n)$. We define an error signal

$$\epsilon(n) = y(n) - \hat{y}(n) \qquad (1.22)$$

where $\hat{y}(n) = \phi^T(n)\hat{\theta}$. This is the same signal as the prediction error in Eq. (1.8). The LMS algorithm defines the cost function as the expected value of the prediction errors as

$$J(n) = E[\epsilon^2(n)] \qquad (1.23)$$

We can write the squared error term as

$$\begin{aligned} \epsilon^2(n) &= (y(n) - \phi^T(n)\hat{\theta})^2 \\ &= y^2(n) - 2y(n)\phi^T\hat{\theta} + \hat{\theta}^T\phi(n)\phi^T(n)\hat{\theta} \end{aligned} \qquad (1.24)$$

We take the expectation and get

$$E[\epsilon^2(n)] = E[y^2(n)] - 2\hat{\theta}^T E[y(n)\phi(n)] + \hat{\theta}^T E[\phi(n)\phi^T(n)]\hat{\theta} \qquad (1.25)$$

We then define the variance $\sigma_y = E[y^2]$ as the mean squared power, and the cross-correlation vector is defined as $p = E[y(n)\phi(n)]$. Then we define the information matrix, which is almost the same matrix as in Section 1.1, as $R = E[\phi(n)\phi^T(n)]$. If the system statistics are stationary, that is, the statistics do not change, the terms σ_y, p, and R, are constants and the cost function, as a function of changing $\hat{\theta}$, will have the shape of a bowl. The cost function $J(n)$ can be written as

$$J(n) = \sigma_y^2 - 2\hat{\theta}^T p + \hat{\theta}^T R\hat{\theta} \qquad (1.26)$$

Once again, as in Eq. (1.4), to find the *optimal* parameter estimate $\hat{\theta}$ to minimize the cost function, we take the partial derivative of the cost function $J(n)$ with respect to $\hat{\theta}$, and determine the value of $\hat{\theta}$ that sets the partial derivative to zero. We can take the partial derivative of $J(n)$ as

$$\frac{\partial J(n)}{\partial \hat{\theta}} = \frac{\partial \sigma_y^2}{\partial \hat{\theta}} - 2\frac{\hat{\theta}^T p}{\partial \hat{\theta}} + \frac{\partial \hat{\theta}^T R\hat{\theta}}{\partial \hat{\theta}} \qquad (1.27)$$

We then compute the partial derivative for each of the terms on the right-hand side of Eq. (1.27). Taking each term separately, we get

$$\frac{\partial \sigma_y^2}{\partial \hat{\theta}} = 0$$

$$2\frac{\hat{\theta}^T p}{\partial \hat{\theta}} = 2p \tag{1.28}$$

$$\frac{\partial \hat{\theta}^T R \hat{\theta}}{\partial \hat{\theta}} = 2R\hat{\theta}$$

Substituting into Eq. (1.27), we get

$$\frac{\partial J(n)}{\partial \hat{\theta}} = -2p + 2R\hat{\theta} = 0 \tag{1.29}$$

Solving for $\hat{\theta}$, we get the solution for the optimal parameter estimate as

$$\theta^* = R^{-1}p \tag{1.30}$$

Equation (1.30) is the well-known Wiener solution. However, the Wiener solution in Eq. (1.30) requires the computation of the inverse of a large matrix R. Notice the similarity between the Wiener solution and the LS solution in Eq. (1.6). Let us say that we want to estimate the expectations in Eq. (1.25); then we would get the average by computing

$$R_{avg} = \left[\frac{1}{N} \sum_{n=1}^{N} \phi(n)\phi^T(n) \right]$$

$$p_{avg} = \left[\frac{1}{N} \sum_{n=1}^{N} \phi(n)y(n) \right] \tag{1.31}$$

Substituting the above values into Eq. (1.30), we get the LS solution given by Eq. (1.6). In essence, the LMS Wiener solution and the LS solution are essentially the same.

In the world of signal processing and in particular adaptive signal processing, the processing speed is very important. Furthermore, the model structure used in adaptive signal processing, especially for communication applications, can have many parameters. It would not be unusual to have 200 terms in the $\phi(n)$ vector, which means the term k in Eq. (1.21) would be $k = 200$. In that case, the R matrix will be 200×200, which would be prohibitively large to take the inverse of in Eq. (1.30). As such, a gradient *steepest descent* method is normally implemented. This is a very common technique throughout the fields of engineering and is very similar to the well-known Newton–Raphson method of finding the zeros and roots of various functions. The steepest descent method is an iterative method. The idea is to start with an initial guess of the parameter values: often one will simply choose zero for the parameter values. In the lexicon of signal processing, one would refer to the parameters as *tap weights*. Then one iteratively adjusts the parameters such that one moves down the cost function along the gradient. Let us say that the current estimate of the parameter vector is $\phi(now)$; then we compute the next value of the parameter vector as

$$\hat{\theta}(next) = \hat{\theta}(now) - \mu g \tag{1.32}$$

where g is the gradient and is given by the derivative of the cost function with respect to the parameter estimation vector, $\hat{\theta}$ as defined in Eq. (1.29). Then, substituting for g in Eq. (1.32), we get

$$\hat{\theta}(next) = \hat{\theta}(now) - \mu 2p - \mu 2R\hat{\theta}(now) \tag{1.33}$$

In recursive form, it is written as

$$\hat{\theta}(n+1) = \hat{\theta}(n) - \mu 2p - \mu 2R\hat{\theta}(n) \tag{1.34}$$

We can also write Eq. (1.34) in the form

$$\hat{\theta}(n+1) = (I - \alpha R)\hat{\theta}(n) - \alpha p \tag{1.35}$$

where $\alpha = 2\mu$. One may recognize, from systems theory, that if the eigenvalues of $(I - \alpha R)$ are less than 1, then the recursion in Eq. (1.35) will converge. This places a limit on the step size of the steepest descent method. We will come back to this point in the next section when we look at the stochastic approximation methods. The effect of step size is an important parameter in machine learning algorithms.

The difficulty in computing the recursion in Eq. (1.34) is the computation of the statistical terms R and p, where R is the information matrix or the autocorrelation matrix, and p is the cross-correlation matrix. Their statistics are often unknown and have to be estimated as we did in Eq. (1.31). However, these estimates are computationally intensive and one has to wait until N data points are collected. Instead, the LMS algorithm proposes that one estimate these matrices based on a single data point at each sampling time, as

$$\hat{R}(n) = \phi(n)\phi^T(n)$$

$$\hat{p}(n) = \phi(n)y(n) \tag{1.36}$$

This is sometimes referred to as the *dirty gradient* method or the *stochastic gradient* method. The idea is that one has to descent in the general direction of the gradient and not exactly along the gradient. Think of yourself walking down a hill; you can either go straight down or, if it is really steep, you may choose to go back and forth traversing the hill, much the same way as a skier. Either way, you end up at the bottom of the hill. Now we substitute the estimates for $\hat{R}$ and $\hat{p}$ given by Eq. (1.36) into the recursive equation given by Eq. (1.34) and we get

$$\hat{\theta}(n+1) = \hat{\theta}(n) + 2\mu\phi(n)y(n) - 2\mu\phi(n)\phi^T(n)\hat{\theta}(n) \tag{1.37}$$

Now we factor out the term $2\mu\phi(n)$ and get the standard LMS recursive algorithm as

$$\hat{\theta}(n+1) = \hat{\theta}(n) + 2\mu\phi(n)(y(n) - \phi^T(n)\hat{\theta}(n)) \tag{1.38}$$

Recall that the term in brackets on the right-hand side, given by $(y(n) - \phi^T(n)\hat{\theta}(n))$, is the prediction error or the innovation. The term $\phi(n)\hat{\theta}(n)$ is the current prediction of the output $y(n)$. If we compare the RLS algorithm in Eq. (1.20) to Eq. (1.38), we see that the update has a similar form. The update to the parameters is the previous estimate plus a matrix vector times the prediction error. In fact, it can be shown that at the stationary point, or the value at which the covarinace matrix update takes the value $P(n+1) = P(n)$ in Eq. (1.20), the LMS algorithm is equivalent to the RLS algorithm for a particular set of parameters.

There is a vast literature on various implementations and convergence results associated with the LMS algorithm, but the key element for this book is that the machine learns the parameters of a preconceived model of the system based on the available experimental data and knowledge of the correct answer given by $y(n)$. The new parameter is the old parameter plus a vector based on the data times the known error in the predicted output.

1.4 Stochastic Approximation

The method of stochastic approximation is an older method of system identification. In fact, it is a method of finding the zeros of a function and is very similar to both the RLS and LMS methods, and it is the fundamental structure for the Q-learning algorithms associated with reinforcement learning and

much of the machine learning literature. The early work in stochastic approximation comes from the work by Robbins and Monro [3] and Wolfowitz [4].
A good textbook on the topic is by Kushner and Yin [5]. Monro formulated the
problem as finding the level at which a continuous function $M(\theta) = \alpha$. Writing
the problem in the form $M(\theta) - \alpha = 0$ converts it into the problem of finding
the zeros of a function. If one knows the gradient of the function, then one can
use the well-known Newton–Raphson method to find the zeros, but in this case
one takes the noise-corrupted measurements of the function at different values
of θ. One then makes small corrections to θ in the estimated direction of zero.

The method of stochastic approximation and the theoretical proofs of stability are used in the proofs of convergence for several fundamental algorithms
in reinforcement learning. Formulating the problem in a similar form to the
previous sections, we get the function $M(\theta) = (y(\theta) - \phi^T \theta) = 0$ and we can
write the prediction error and the error in getting to zero as $\epsilon = (y - \phi^T \theta)$.
The stochastic approximation algorithm is

$$\theta(n + 1) = \theta(n) - a_n(y - \phi^T \theta) \tag{1.39}$$

where a_n is a variable step size that goes to zero such that

$$0 < \sum_1^\infty a_n^2 = A < \infty \tag{1.40}$$

References

[1] L. Ljung and T. Soderstrom, *Theory and Practice of Recursive Identification.* Cambridge,
    Massachusetts: The MIT Press, 1983.

[2] B. Farhang-Boroujeny, *Adaptive Filters: Theory and Applications.* New York: John Wiley &
    Sons, Inc., 1998.

[3] H. Robbins and S. Monro, "A stochastic approximation method," *Annals of Mathematical
    Statistics*, vol. 22, no. 3, pp. 400–407, 1951.

[4] J. Wolfowitz, "On the stochastic approximation method of robbins and monro," *Annals of
    Mathematical Statistics*, vol. 23, no. 3, pp. 457–461, 1952.

[5] H. J. Kushner and G. G. Yin, *Recursive Approximation and Recursive Algorithms and Appli-
    cations.* New York: Springer-Verlag, 2nd ed., 2003.

Chapter 2
Single-Agent Reinforcement Learning

The objective of this chapter is to introduce the reader to reinforcement learning. A good introductory book on the topic is Reference 1 and we will follow their notation. The goal of reinforcement learning is to maximize a reward. The interesting aspect of reinforcement learning, as well as unsupervised learning methods, is the choice of rewards. In this chapter, we will discuss some of the fundamental ideas in reinforcement learning which we will refer to in the rest of the book. We will start with the simple n-armed bandit problem and then present ideas on the meaning of the "value" function.

2.1 Introduction

Reinforcement learning is learning to map situations to actions so as to maximize a numerical reward [1]. Without knowing which actions to take, the learner must discover which actions yield the most reward by trying them. Actions may affect not only the immediate reward but also the next situation and all subsequent rewards [1]. Different from supervised learning, which is learning from examples provided by a knowledgeable external supervisor, reinforcement learning is used for learning from interaction [1]. Since it is often impractical to obtain examples of desired behavior that are both correct and representative of all the situations, the learner must be able to learn from its

Multi-Agent Machine Learning: A Reinforcement Approach, First Edition. Howard M. Schwartz.
© 2014 John Wiley & Sons, Inc. Published 2014 by John Wiley & Sons, Inc.

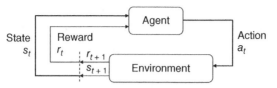

Fig. 2-1. Agent–environment interaction in reinforcement learning. Reproduced from [1], with permission of MIT Press.

own experience [1]. Therefore, the reinforcement learning problem is a problem of learning from interaction to achieve a goal.

The learner is called the *agent* or the *player* and the outside which the agent interacts with is called the *environment*. The agent chooses actions to maximize the rewards presented by the environment. Suppose we have a sequence of discrete time steps $t = 0, 1, 2, 3, \ldots$. At each time step t, the agent receives a state s_t from the environment. We define a_t as the action the agent takes at time t. At the next time step, as a consequence of its action a_t, the agent receives a numerical reward $r_{t+1} \in \Re$ and moves to a new state s_{t+1}, as shown in Fig. 2-1. At each time step, the agent implements a mapping from states to probabilities of selecting each possible action [1]. This mapping is called the *agent's policy* and is denoted as π_t, and $\pi_t(s, a)$ is the probability of $a_t = a$ at $s_t = s$. Reinforcement learning methods specify how the agent changes its policy as a result of its experience to maximize the total amount of reward it receives over the long run [1].

A reinforcement learning problem can be studied under the framework of stochastic games [2]. Such a framework contains two simpler frameworks: Markov decision processes (MDPs), and matrix games [2]. MDPs involve a single agent and multiple states, while matrix games include multiple agents and a single state. Combining MDPs and matrix games, stochastic games are considered as reinforcement learning problems with multiple agents and multiple states.

2.2 *n*-Armed Bandit Problem

The *n*-armed bandit problem is taken from playing slot machines. The idea is that one is playing a slot machine with n arms. Each arm will give you a different reward, or a different probability of winning or losing. The idea is to determine which of the n arms will give you the greatest reward. Therefore, there are n actions that one can take, each of the actions representing one of the arms that you can pull. The action that gives the greatest reward is referred to as the *greedy* reward. How would one learn which is the best arm to pull to get the greatest expected reward? You would probably try and pull each arm many times and try to compute a running average of which arm seemed to give you the most reward. We will define the value of each arm as the expected reward for the arm, or action. In this case, pulling a given arm is the same as choosing an action. Let us estimate the value of an action at trial t as

$$Q_t(a) = \frac{r_1 + r_2 + \cdots + r_k}{k} \tag{2.1}$$

where r_i is the reward of choosing arm a on the ith time step. However, there were t actions played in total, but only k of those times was the arm or action a played. We will denote the actual value of the arm or action a as $Q^*(a)$. The rule for choosing which action to take is to always choose the *greedy* action except for the relatively small probability of ϵ of choosing a random action. The *greedy* actions are taken based on which action has the highest expected reward at that particular time step. However, sometimes one should explore to find out if some other choice of action would be better. This random selection of some action that is not the *greedy* one is referred to as *exploration*. Within machine learning, there is a kind of tension between how much *exploration* there should be and how much *exploitation* there should be.

Let us take a simple example of the 10-armed bandit. We start by assigning a random reward for each of the 10 arms from a normal random distribution with mean zero and variance 1. We take 10 numbers from the distribution $N(0, 1)$. These 10 numbers will represent the true value or the expected reward for each of the 10 arms. We get the true reward for each arm as

$$Q^*(a) = [-0.4 \quad 1.3 \quad 0.04 \quad 0.53 \quad -0.15 \quad -1.01 \quad 0.2 \quad 1.48 \quad 0.36 \quad -0.5] \tag{2.2}$$

We know that the best choice of action is action 8, with $Q^*(8) = 1.48$. This is the correct answer that we want the machine to learn. We also set the *exploration* variable to $\epsilon = 0.2$. This means that there is a 20% chance that on any given choice the machine will randomly choose an action without regard to the estimate of the reward.

We start the learning process by guessing at the expected reward for each arm by choosing our initial guess from the same normal distribution that we chose the true rewards from, and the true rewards are given in Eq. (2.2). We then get our first estimate of the true reward as

$$Q_{est}(0) = [0.05 \quad 0.86 \quad -0.96 \quad 0.73 \quad 1.98 \quad -1.19 \quad -0.66 \quad 0.82 \quad 1.97 \quad -0.13] \tag{2.3}$$

Based on the initial estimate, we should choose action 5, which is estimated by Eq. (2.3) as $Q_{est}(5) = 1.98$, but the true value of action 5 is really $Q^*(5) = -0.15$, and therefore the initial *greedy* choice we took is a very poor choice. The machine chooses action 5, and the resulting reward is selected from a random distribution of $N(-0.15, 1)$, which has a mean of -0.15 and a variance of 1. The machine gets the reward $Q_1(5) = -0.76$. The machine then updates the $Q_{est}(a)$ table or vector according to Eq. (2.1) as

$$Q_{est}(1) = [0.05 \quad 0.86 \quad -0.96 \quad 0.73 \quad -0.76 \quad -1.19 \quad -0.66 \quad 0.82 \quad 1.97 \quad -0.13]$$
(2.4)

The machine then chooses another arm based on its most recent estimate of the rewards given by Eq. (2.4). The machine once again makes the *greedy* choice and chooses action 9 because $Q_{est}(9) = 1.97$. The machine chooses action $a = 9$, and the machine gets the reward $Q_2(9) = 1.50$. We then get the estimate of $Q_{est}(2)$ as

$$Q_{est}(2) = [0.05 \quad 0.86 \quad -0.96 \quad 0.73 \quad -0.76 \quad -1.19 \quad -0.66 \quad 0.82 \quad 1.5 \quad -0.13]$$
(2.5)

We repeat this process and, on trial number 6, the machine explores and arbitrarily chooses action 6, which we know is not a good choice, because from Eq. (2.2) the true average reward for action $a = 6$ is $Q^*(6) = -1.01$. The reward the machine gets for choosing action $a = 6$ on the sixth trial is $r = 1.44$. The learning process continues until the 18th trial, when it finally recognizes action $a = 8$ as the best choice. The results of implementing the 10-armed bandit problem for different values of ϵ is illustrated in Fig. 2-2.

2.3 The Learning Structure

In this section, we will formulate the basic structure for reinforcement learning. We will take the case of a single agent. The agent, or robot, is in a particular state and in that state it can take one of a number of actions. In some cases, there exists a particular action that is in some sense optimal for that particular state. We will sometimes refer to that action as the *greedy* action and the set of actions the agent should *optimally* take in each state is the action *policy* or the *strategy*. The terms *policy* and *strategy* are sometimes used interchangeably and sometimes they may have slightly different meanings: it depends on the context. If in a particular state the agent should choose a particular action, we will refer to that as a *pure strategy*, whereas if the agent should choose an action with a particular probability, then we refer to that as a *distributed policy*, or a mixed strategy. For example, if you are playing the rock-paper-scissors game, then the optimal strategy is to take each action with a probability of $1/3$, but if

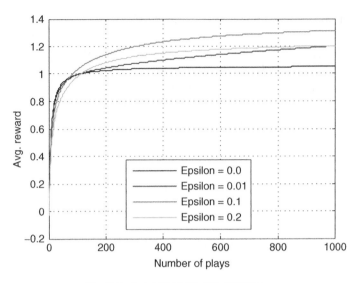

Fig. 2-2. Armed bandit with varying ϵ.

you knew in advance that your opponent will always choose Rock, then your *optimal* strategy or policy is to choose the pure strategy of paper.

We will define the actions as $a_t \in A(s_t)$. The action is a member of an action set for the given state. The possible actions an agent can take depend on the state the agent is in. Furthermore, there is a defined number of possible states. Here we have $s_t \in S$. If it is currently time t, our goal will be to maximize future rewards, given by

$$R_t = r_{t+1} + r_{t+2} + \cdots + r_T \tag{2.6}$$

where T is the terminal time. This type of reward function works well when we have episodic games that have a well-defined terminal time. However, in many cases we do not have a terminal time and the game continues indefinitely, such as a regulator in control applications. In those cases, without well-defined terminal times, one typically uses discounted future rewards. In this case, the reward is defined as

$$R_t = r_{t+1} + \gamma r_{t+2} + \gamma^2 r_{t+3} \cdots = \sum_{k=0}^{\infty} \gamma^k r_{t+k+1} \tag{2.7}$$

where γ is the discount factor given as $0 \leq \gamma \leq 1$. If γ is close to zero, then we refer to the algorithm as *myopic*, but if γ is close to 1, then the algorithm is maximizing future rewards.

In reinforcement learning, we want the current state to be a good basis for predicting the future. Such systems are called *MDPs*. A decision system is said to have the Markov property when the current state is all one needs to know to make decisions about what to do next and into the future. What to do in the next step does not depend on what happened in the past, but just the state the system is currently in. For example, a game of checkers depends only on the current state of the game and not how one got to that position. A more engineering example is that the future flight of a ball depends only on its current position and velocity vector; it does not depend on how it got there. In the Markov world, we can define the Markov property as the probability of receiving a particular reward, and changing to another state is completely defined by the current state and action as $Pr\{s_{t+1} = s', r_{t+1} = r | s_t, a_t\}$; whereas in the non-Markovian world the same probability would be written as $Pr\{s_{t+1} = s', r_{t+1} = r | s_t, a_t, s_{t-1}, a_{t-1}, \ldots s_0, a_0\}$. We can define the probability of taking action a and transitioning from state s to the next state s' as $P^a_{ss'} = Pr\{s_{t+1} = s' | s_t = s, a_t = a\}$. Furthermore, we can define the expected reward as $R^a_{ss'} = E\{r_{t+1} | s_t = s, a_t = a, s_{t+1} = s'\}$.

2.4 The Value Function

The value function, in the framework of reinforcement learning defines *how good* a particular state is. The measure of how good a given state is, is based on the expected future rewards that will be achieved from that state. The players in the game follow a strategy $\pi(s, a)$, which is the probability of taking action a when in state s. The value of a state s is the expected future return of following policy $\pi(s, a)$ starting from s, and can be written as

$$V^\pi(s) = E_\pi\{R_t | s_t = s\} = E_\pi\left\{\sum_{k=0}^{\infty} \gamma^k r_{t+k+1} | s_t = s\right\} \tag{2.8}$$

We also define the action value function for policy or strategy, π:

$$Q^\pi(s, a) = E_\pi\{R_t | s_t = s, a_t = a\}$$

$$= E_\pi\left\{\sum_{k=0}^{\infty} \gamma^k r_{t+k+1} | s_t = s, a_t = a\right\} \tag{2.9}$$

The term $Q^\pi(s, a)$ is a little different from $V^\pi(s)$. The term $Q^\pi(s, a)$ is the expected return if we choose action a in state s and then follow policy or strategy π afterwards. We can write the value function as

$$V^\pi(s) = E_\pi\{R_t | s_t = s\} = E_\pi\left\{\sum_{k=0}^{\infty} \gamma^k r_{t+k+1} | s_t = s\right\} \tag{2.10}$$

$$= E_\pi\left\{r_{t+1} + \sum_{k=1}^{\infty} \gamma^k r_{t+k+1} | s_t = s\right\} \tag{2.11}$$

$$= E_\pi\left\{r_{t+1} + \gamma \sum_{k=0}^{\infty} \gamma^k r_{t+k+2} | s_t = s\right\}. \tag{2.12}$$

We can then write this in the form

$$V^\pi(s) = \sum_a \pi(s,a) \sum_{s'} P_{ss'}^a \left(R_{ss'}^a + \gamma E_\pi\left\{\sum_{k=0}^{\infty} \gamma^k r_{t+k+2} | s_t = s\right\}\right)$$

$$= \sum_a \pi(s,a) \sum_{s'} P_{ss'}^a (R_{ss'}^a + \gamma V^\pi(s')). \tag{2.13}$$

The first sum on the right-hand side of Eq. (2.13) represents the sum over all the possible actions in state s, where $\pi(s,a)$ is the probability of taking a particular action. The second sum in Eq. (2.13) is taken over all the possible next states. Recall that $P_{ss'}^a$ is the probability of transitioning from state s to state s' given one has chosen action a. The last term in brackets on the right represents the immediate reward $R_{ss'}^a$ plus the future discounted rewards from the next state onwards, given by $V^\pi(s')$. Equation (2.13) is known as the *Bellman equation* for $V^\pi(s)$. Similarly, we can write the state-action value function as

$$Q(s,a) = \sum_{s'} P_{ss'}^a (R_{ss'}^a + \gamma V^\pi(s')) \tag{2.14}$$

2.5 The Optimal Value Functions

We define the optimal state-value function as

$$V^*(s) = \max_\pi V^\pi(s) \ \forall s \in S \tag{2.15}$$

and, similarly, the optimal state-action value function is given as

$$Q^*(s,a) = \max_\pi Q^\pi(s,a) \tag{2.16}$$

As such, we are searching for the policy, or the choice of actions that will give the greatest reward. Then $Q(s, a)$ can be written as

$$Q(s, a) = E\left\{r_{t+1} + \gamma \sum_{k=0}^{\infty} \gamma^k r_{t+k+2} | s_t = s, a_t = a\right\} \qquad (2.17)$$

Then the optimal value becomes

$$Q^*(s, a) = E\{r_{t+1} + \gamma V^*(s_{t+1}) | s_t = s, a_t = a\} \qquad (2.18)$$

The term $Q^*(s, a)$ represents one getting an immediate reward of r_{t+1} for taking action a and then following the optimal policy thereafter. We can write the value function in the following form:

$$V^*(s) = \max_{a \in A(s)} Q^{\pi*}(s) \qquad (2.19)$$

The value function and the state-action function look to the future. They refer to the expected future rewards. Therefore, if we know the maximum future rewards from the next state onwards, then all we have to do is to choose the best action right now. We can write Eq. (2.19) as

$$\begin{aligned}
V^*(s) &= \max_{a \in A(s)} Q^{\pi*}(s) \\
&= \max_{a \in A(s)} E_{\pi*}\{R_t | s_t = s, a_t = a\} \\
&= \max_{a \in A(s)} E_{\pi*}\left\{\sum_{k=0}^{\infty} \gamma^k r_{t+k+1} | s_t = s, a_t = a\right\} \\
&= \max_{a \in A(s)} E_{\pi*}\left\{r_{t+1} + \gamma \sum_{k=0}^{\infty} \gamma^k r_{t+k+2} | s_t = s, a_t = a\right\} \\
&= \max_{a \in A(s)} E_{\pi*}\{r_{t+1} + \gamma V^*(s_{t+1}) | s_t = s, a_t = a\} \\
&= \max_{a \in A(s)} \sum_{s'} P_{ss'}^a (R_{ss'}^a + \gamma V^*(s_{t+1})).
\end{aligned} \qquad (2.20)$$

Similarly we can write

$$Q^*(s, a) = E\left\{r_{t+1} + \gamma \max_{a'} Q^*(s_{t+1}, a') | s_t = s, a_t = a\right\} \qquad (2.21)$$

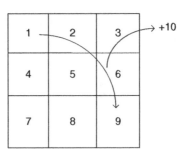

Fig. 2-3. Example of the grid world.

2.5.1   The Grid World Example

We will take an example of a 3×3 grid, as shown in Fig. 2-3. The robot or agent can be in one of the nine cells at any starting time. It can then move in one of four directions. The actions then are taken from the set $a \in [up, down, right, left]$. When the robot or learning agent moves to cell 1, it then immediately moves to cell 9 and gets a reward of $r = +10$. If the agent hits a wall, it remains in its current cell and gets a negative reward (punishment) of $r = -1$. We will use Eq. (2.20) to derive the nine equations and nine unknowns. We will solve Eq. (2.20) with the following parameters: the discount factor for future rewards $\gamma = 0.9$, and the action selection strategy is $\pi(s, a) = 0.25$; in other words, the agent will arbitrarily choose to go either up, down, left, or right with equal probability. There are four possible actions to transition from a particular state. The only reward is when the agent gets to state 1 and then transitions with 100% probability to state 9 and receives a reward of +10. The rewards are all zero except for $r_{19}^{up} = 10, r_{19}^{down} = 10, r_{19}^{left} = 10, r_{19}^{right} = 10$ and, when the agent hits the wall, $r_{22}^{up} = -1, r_{33}^{up} = -1, r_{33}^{right} = -1, r_{44}^{left} = -1, r_{66}^{right} = -1, r_{77}^{down} = -1, r_{77}^{left} = -1, r_{99}^{down} = -1, r_{99}^{right} = -1$. The transition probabilities are as follows:

$$P_{19}^{up} = 1, \quad P_{19}^{down} = 1, \quad P_{19}^{left} = 1, \quad P_{19}^{right} = 1$$

$$P_{22}^{up} = 1, \quad P_{25}^{down} = 1, \quad P_{21}^{left} = 1, \quad P_{23}^{right} = 1$$

$$P_{33}^{up} = 1, \quad P_{36}^{down} = 1, \quad P_{32}^{left} = 1, \quad P_{33}^{right} = 1$$

$$P_{41}^{up} = 1, \quad P_{47}^{down} = 1, \quad P_{44}^{left} = 1, \quad P_{45}^{right} = 1$$

$$P_{52}^{up} = 1, \quad P_{58}^{down} = 1, \quad P_{54}^{left} = 1, \quad P_{56}^{right} = 1$$

$$P_{63}^{up} = 1, \quad P_{69}^{down} = 1, \quad P_{65}^{left} = 1, \quad P_{66}^{right} = 1$$

$$P_{74}^{up} = 1, \quad P_{77}^{down} = 1, \quad P_{77}^{left} = 1, \quad P_{78}^{right} = 1$$

$$P^{up}_{85} = 1, \quad P^{down}_{88} = 1, \quad P^{left}_{87} = 1, \quad P^{right}_{89} = 1$$
$$P^{up}_{96} = 1, \quad P^{down}_{99} = 1, \quad P^{left}_{98} = 1, \quad P^{right}_{99} = 1. \tag{2.22}$$

We can now substitute these into Eq. (2.13) to get the nine equations and nine unknowns for each state. Let us specifically write out the first equation for the first state, recalling that $\pi(s, a) = 0.25$, representing equal probability for the taking of each action and $\gamma = 0.9$. Writing out the $V^{\pi}(1)$ equation is a special case because, regardless of which action we take, the we always move to state 9 and receive a reward of $r^a_{19} = +10$. Therefore, we get

$$V^{\pi}(1) = \sum_{a=1}^{4}(0.25) \left(\sum_{9} P^a_{19} \left((r^a_{19} + \gamma V^{\pi}(9)) \right) \right) \tag{2.23}$$

The term

$$\sum_{9} P^a_{19}(r^a_{19} + \gamma V^{\pi}(9)) = r^a_{19} + 0.9 V^{\pi}(9) \tag{2.24}$$

dropping the superscript on V^{π} for ease of writing, gives the first equation as

$$V(1) = 10 + 0.9 V(9) \tag{2.25}$$

Now we will go through the same process for state 2. In this case, we do not automatically jump to another state but we can move to one of four other states. Recall that hitting a wall will give us a negative reward of -1. Then we can write

$$V(2) = \sum_{a=1}^{4}(0.25) \left(\sum_{s'} P^a_{2s'}(r^a_{2s'} + \gamma v(s')) \right)$$
$$= 0.25 \left(\sum_{s'} P^{up}_{2s'}(r^{up}_{2s'} + \gamma V(s')) \right) + 0.25 \left(\sum_{s'} P^{down}_{2s'}(r^{down}_{2s'} + \gamma V(s')) \right)$$
$$+ 0.25 \left(\sum_{s'} P^{left}_{2s'}(r^{left}_{2s'} + \gamma V(s')) \right) + 0.25 \left(\sum_{s'} P^{right}_{2s'}(r^{right}_{2s'} + \gamma V(s')) \right). \tag{2.26}$$

If we examine the probabilities on the right of Eq. (2.26), we note that they take a value of either 1 or 0. For example, if we look at the term $P^{up}_{2s'}$, it only takes a value for $P^{up}_{22} = 1$, and all other state transitions for $P^{up} = 0$. Furthermore, if the agent tries to go up, it hits the wall and gets a reward of -1. We can now rewrite equation 2.26 as

$$V(2) = 0.25(-1 + 0.9V(2)) + 0.25(0 + 0.9V(5)) + 0.25(0 + 0.9V(1))$$
$$+ 0.25(0 + 0.9V(3))$$
$$= 0.225V(1) + 0.225V(2) + 0.225V(3) + 0.225V(5) - 0.25.$$

Just for some further practice, we will write out the equation for $V(3)$. In this case, the agent will hit the wall if it moves up or right, and will receive a reward of -1. Then we get

$$V(3) = 0.25(-1 + 0.9V(3)) + 0.25(0 + 0.9V(6)) + 0.25(-1 + 0.9V(3))$$
$$+ 0.25(0 + 0.9V(2))$$
$$= 0.225V(2) + 0.45V(3) + 0.225V(6) - 0.5.$$

We continue going through the process of determining the equations for each state, and we then write out the equations in matrix form as $AV = B$, where A is given as

$$A = \begin{bmatrix}
1 & 0 & 0 & 0 & 0 & 0 & 0 & 0 & -0.9 \\
-0.225 & 0.775 & -0.225 & 0 & -0.225 & 0 & 0 & 0 & 0 \\
0 & -0.225 & 0.55 & 0 & 0 & -0.225 & 0 & 0 & 0 \\
-0.225 & 0 & 0 & 0.775 & -0.225 & 0 & -0.225 & 0 & 0 \\
0 & -0.225 & 0 & -0.225 & 1 & -0.225 & 0 & -0.225 & 0 \\
0 & 0 & -0.225 & 0 & -0.225 & 0.775 & 0 & 0 & -0.225 \\
0 & 0 & 0 & -0.225 & 0 & 0 & 0.55 & -0.225 & 0 \\
0 & 0 & 0 & 0 & -0.225 & 0 & -0.225 & 0.775 & -0.225 \\
0 & 0 & 0 & 0 & 0 & -0.225 & 0 & -0.225 & 0.55
\end{bmatrix}$$

and B is given by

$$B^T = \begin{bmatrix} 10 & -0.25 & -0.5 & -0.25 & 0 & -0.25 & -0.5 & -0.25 & -0.5 \end{bmatrix}$$

Solving for this set of nine equations and nine unknowns, we get the true value of each state as

$$V^T = \begin{bmatrix} 8.85 & 2.5 & -0.07 & 2.5 & 0.92 & -0.44 & -0.07 & -0.44 & -1.27 \end{bmatrix} \quad (2.27)$$

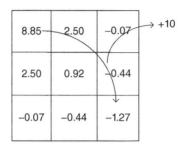

Fig. 2-4. Values for each of the states.

Recall that the value function for each state is for the future discounted rewards. Therefore, state 1 does not get a value of 10 but a value of 8.85 because we do not know where the agent goes from state 9 onwards. We show the value of each state in the grid in Fig. 2-4.

This is an important result as we delve into machine learning algorithms. In some way, we need to learn the value of the state. Furthermore, even for this relatively simple stochastic game, look at the amount of information we needed to know to solve for the value of the state. We needed to know all the transition probabilities from state to state, based on what action was taken, and in this case it was relatively easy; the transition probabilities were either 1 or 0. We have to know all possible actions in each state, and we need to know the rewards a priori.

The goal of machine learning is to try to automatically learn what would be the best path to take, given any initial state, such that you collect the greatest reward. You and I can look at this game and almost immediately recognize the best path to take, but what if you knew nothing about the game. How would you learn what the best path would be? You would try out some steps until you eventually hit state 1 and collected your big reward. This is what we want the agent to eventually learn how to do.

2.6 Markov Decision Processes

An MDP [3] is a tuple (S, A, T, γ, R), where S is the state space, A is the action space, $T : S \times A \times S \rightarrow [0, 1]$ is the transition function, $\gamma \in [0, 1]$ is the discount factor, and $R : S \times A \times S \rightarrow \mathbb{R}$ is the reward function. The transition function denotes a probability distribution over next states given the current state and action such that

$$\sum_{s' \in S} T(s, a, s') = 1 \quad \forall s \in S, \quad \forall a \in A \qquad (2.28)$$

where s' represents a possible state at the next time step. The reward function denotes the received reward at the next state given the current action and the current state. An MDP has the following Markov property: the player's next state and reward depend only on the player's current state and action. A player's policy $\pi : S \to A$ is defined as a probability distribution over the player's actions from a given state. A player's policy $\pi(s, a)$ satisfies

$$\sum_{a \in A} \pi(s, a) = 1 \quad \forall s \in S \tag{2.29}$$

For any MDP, there exists a deterministic optimal policy for the player, where $\pi^*(s, a) \in \{0, 1\}$ [4]. The goal of a player in an MDP is to maximize the expected long-term reward. In order to evaluate a player's policy, we have the following concept of the state-value function: The value of a state s (or the state-value function) under a policy π is defined as the expected return when the player starts at state s and follows a policy π thereafter. Then the state-value function $V^\pi(s)$ becomes

$$V^\pi(s) = E_\pi \left\{ \sum_{k=0}^{T} \gamma^k r_{k+t+1} | s_k = s \right\} \tag{2.30}$$

where T is a final time step, t is the current time step, r_{k+t+1} is the received immediate reward at the time step $k + t + 1$, and $\gamma \in [0, 1]$ is a discount factor. In (2.30), we have $T \to \infty$ if the task is an infinite-horizon task such that the task will run over an infinite period. If the task is episodic, T is defined as the terminal time when each episode is terminated at the time step T. Then we call the state where each episode ends as the *terminal state* s_T. In a terminal state, the state-value function is always zero such that $V(s_T) = 0 \ \forall s_T \in S$. An optimal policy π^* will maximize the player's discounted future reward for all states such that

$$V^*(s) \geq V^\pi(s) \quad \forall \pi, \forall s \in S \tag{2.31}$$

The state-value function under a policy in (2.30) can be rewritten as a recursive equation called the *Bellman equation* [5] as follows:

$$V^\pi(s) = \sum_{a \in A} \pi(s, a) \sum_{s' \in S} T(s, a, s')(R(s, a, s') + \gamma V^\pi(s')) \tag{2.32}$$

where $T(s, a, s') = Pr\{s_{k+1} = s' | s_k = s, a_k = a\}$ is the probability of the next state being $s_{k+1} = s'$ given the current state $s_k = s$ and action $a_k = a$ at time step k, and $R(s, a, s') = E\{r_{k+1} | s_k = s, a_k = a, s_{k+1} = s'\}$ is the expected immediate reward received at state s' given the current state s and action a. If the player starts at state s and follows the optimal policy π^* thereafter, we have the optimal state-value function denoted by $V^*(s)$. The optimal state-value function $V^*(s)$ is also called the *Bellman optimality equation*, where

$$V^*(s) = \max_{a \in A} \sum_{s' \in S} T(s, a, s')(R(s, a, s') + \gamma V^*(s')) \qquad (2.33)$$

We can also define the action-value function as the expected return of choosing a particular action a at state s and following a policy π thereafter. The action-value function $Q^\pi(s, a)$ is given as

$$Q^\pi(s, a) = \sum_{s' \in S} T(s, a, s')(R(s, a, s') + \gamma V^\pi(s')) \qquad (2.34)$$

If the player chooses action a at state s and follows the optimal policy π^* thereafter, the action-value function becomes the optimal action-value function $Q^*(s, a)$, where

$$Q^*(s, a) = \sum_{s' \in S} T(s, a, s')(R(s, a, s') + \gamma V^*(s')) \qquad (2.35)$$

In a terminal state s_T, the action-value function is always zero such that $Q(s_T, a) = 0 \ \forall \ s_T \in S$.

2.7 Learning Value Functions

In the previous section, we computed the value function for each state by solving nine equations and nine unknowns. Also we saw in the previous section that just writing out the equations can be tedious even for the relatively simple example that we had. In this section, we will present the iterative computer algorithm that will solve the equation for the value functions. This type of approach is known more widely as *dynamic programming*. Given that we already know the transition probabilities, the action strategies, and the rewards, we can compute the value of the state as in Eq. (2.13). Recall that we wrote the equation for the value function in the form $AV = B$, and we can also write it as $AV - B = 0$, which is solved for as $V = A^{-1}B$, and most of us learned to solve this equation using Gaussian elimination. Let us say we just write an algorithm to find the zero of $AV - B = 0$. Any dirty type gradient algorithm can work. Most people who have done technical work have done this type of thing. Sometimes they tried to find the argument of a function by trial and error. The idea is to solve Eq. (2.36) iteratively. We will not write out the set of equations, but we will write a machine learning algorithm to find it. Furthermore, we will construct this learning algorithm as a gradient search algorithm to find the zero of a function. In more formal terms, we are

---

Algorithm 2.1 Value iteration algorithm

Initialize $V(s) = 0$ for all $s \in S$
repeat
 Set $\Delta = 0$
 For each $s \in S$:
 $v \leftarrow V(s)$
 $V(s) \leftarrow \sum_a \pi(s, a) \sum_{s'} P^a_{ss'} \left(R^a_{ss'} + \gamma V_k(s') \right)$
 $\Delta = \max(\Delta, |v - V(s)|)$
until $\Delta < \theta$ for all $s \in S$ (θ is a small positive number)

---

searching for the stationary point of a recursive algorithm, which is similar to stochastic approximation.

The algorithm works as follows: we initialize the value function to zero as $V(s) = 0$ $\forall s$. We then compute Eq. (2.13) and repeating it for convenience as

$$V_{k+1}(s) = \sum_a \pi(s, a) \sum_{s'} P^a_{ss'}(R^a_{ss'} + \gamma V_k(s')) \qquad (2.36)$$

This gives us the first guess at the value of each state. Recall, we must know $\pi(s, a)$, $P^a_{ss'}$, and $R^a_{ss'}$ in advance. We then compute the difference $\Delta = V_{k+1} - V_k$. Once this difference becomes small enough, we stop the recursion given by Eq. (2.36). Once we get to $V_{k+1} - V_k \Rightarrow 0$, then the algorithm given by the recursion in (2.36) has reached a stationary point. The value iteration algorithm is given in Algorithm 2.1. We set $\Delta = 0.001$, and the algorithm executes 40 iterations before it exits with the result

$$V^T = \begin{bmatrix} 8.85 & 2.5 & -0.07 & 2.5 & 0.92 & -0.44 & -0.07 & -0.44 & -1.27 \end{bmatrix} \quad (2.37)$$

which is the same result that we got with the exact solution in Eq. (2.27).

2.8 Policy Iteration

In the previous section, we presented a recursive algorithm to solve the n equations and n unknown problem to compute the value of each state. However, we used the known state transition matrices $P^a_{ss'}$, the known rewards $R^a_{ss'}$, and the known action strategies $\pi(s, a)$ to do the computations. However, can we determine whether a better strategy would give us a higher reward. If we searched over all the possible strategies, could we find a strategy that would yield a higher reward and what would the resulting value of the states be under that new strategy? In this section, we present the *policy iteration* algorithm. This algorithm will compute both the value of the state and the *optimal* action strategy. Recall, we wrote out the state-action value function in Eq. (2.14) as

$$Q(s, a) = \sum_{s'} P^a_{ss'} (R^a_{ss'} + \gamma V(s'))$$ (2.38)

The goal is to search for the policy that maximizes $Q(s, a)$, as

$$Q^*(s, a) = \max_a \sum_{s'} P^a_{ss'} (R^a_{ss'} + \gamma V(s'))$$ (2.39)

The algorithm has two phases. In the first phase, the value of each state is computed as was done in the value iteration algorithm. In the next phase, we compute Eq. (2.39) and search for the maximizing action. If more than one action yields the equivalent maximizing result, then we assign the probability of each action equivalently. For example, if two actions give the same maximizing result, then each action gets a probability of 50%. We implement the *policy iteration* Algorithm 2.2 on the previous grid game. This time we get

---

Algorithm 2.2 Policy iteration algorithm

1: Initialize $V(s) = 0$ for all $s \in S$

2: Policy Evaluation
3: **repeat**
4:     $\Delta = 0$
5:     For each $s \in S$:
6:         $v \leftarrow V(s)$
7:         $V(s) \leftarrow \sum_a \pi(s, a) \sum_{s'} P^a_{ss'} \left(R^a_{ss'} + \gamma V_k(s') \right)$
8:         $\Delta = \max(\Delta, |v - V(s)|)$
9: **until** $\Delta < \theta$ for all $s \in S$ (θ is a small positive number)

10: Policy Improvement
11: **repeat**
12:     For each $s \in S$:
13:         $b \leftarrow \pi(s)$
14:         $\pi(s) \leftarrow \max_a \sum_{s'} P^a_{ss'} \left(R^a_{ss'} + \gamma V_k(s') \right)$
15:         If $b \neq \pi(s)$ Then go back to policy evaluation
16: **until** $b = \pi(s)$

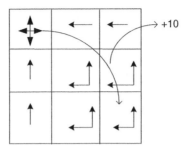

Fig. 2-5. Resulting optimal policies.

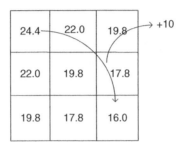

Fig. 2-6. Resulting state values based on the optimal policies.

a different strategy and the value function increases substantially. Figure 2-5 illustrates the resulting strategies for each state and the associated probability. For example, state 1 has four directions and equal probability of choosing any action because the next state is always state 9 regardless of the chosen action; whereas in state 2 the optimal action is to always choose with 100% probability to take action *left* and move to state 1. When in state 5, the *up* and *left* actions each have equal probability of 50% (Fig. 2-6). Based on this new strategy, the value of each state can then be computed as

$$V^T = \begin{bmatrix} 24.4 & 22.0 & 19.8 & 22.0 & 19.8 & 17.8 & 19.8 & 17.8 & 16.0 \end{bmatrix}$$

Notice the dramatic improvement in the value of each state in comparison to the values in (2.37). This is because we have now computed the value of the state based on the *optimal* action strategy.

2.9 Temporal Difference Learning

The idea of reinforcement learning is to learn the best policy and actions. In Section 2.7 we iteratively learnt, or more accurately we calculated, the value of the state on the basis of the knowledge of the action strategies $\pi(s, a)$, the state transition probabilities $P^a_{ss'}$, and the rewards $R^a_{ss'}$. In Section 2.8, we presented the policy iteration algorithm that searched for the *optimal* policy and we saw how the optimal policy would improve the value of the states. However, in this case we had to know the state transition probabilities $P^a_{ss'}$ and the rewards $R^a_{ss'}$. In this section, we will present a machine learning algorithm that will learn the value of the state based on just observing the rewards from the environment.

Let us return to the original definition of the value of the state as the expected future rewards as we wrote in (2.13):

$$V^{\pi}(s) = E_{\pi}\{R_t|s_t = s\} = E_{\pi}\left\{\sum_{k=0}^{\infty}\gamma^k r_{t+k+1}|s_t = s\right\} \tag{2.40}$$

$$= E_{\pi}\left\{r_{t+1} + \sum_{k=1}^{\infty}\gamma^k r_{t+k+1}|s_t = s\right\} \tag{2.41}$$

$$= E_{\pi}\left\{r_{t+1} + \gamma\sum_{k=0}^{\infty}\gamma^k r_{t+k+2}|s_t = s\right\}. \tag{2.42}$$

We can then write this in the form

$$V^{\pi}(s) = r_{t+1} + \gamma V^{\pi}(s') \tag{2.43}$$

Notice that we have identified the value of the state with the action strategy π. If we know the transition probabilities and the rewards, then we can solve for the value of the state by either deriving the n equations and n unknowns as we did in Section 2.5.1 or by dynamic programming as we did in Section 2.7. However, in this case we do not have the luxury of knowing the transition probabilities $P^a_{ss'}$ nor the rewards $R^a_{ss'}$. Therefore, we will use another search method that we are familiar with from the least mean square (LMS) approach. Let us begin with an initial estimate of the value function for each state. We will often, in practice, initialize the value function to zero. Let us rewrite Eq. (2.43) in the form

$$\epsilon = r_{t+1} + \gamma V(s') - V(s) \tag{2.44}$$

Now we will update our estimate of the value of the state as

$$V_{k+1}(s) = V_k(s) + \alpha[r_{t+1} + \gamma V(s') - V(s)] \tag{2.45}$$

This is the same structure that we saw in the case of the LMS algorithm defined in Eq. (1.38). The update is the previous value plus learning rate α (or step size) times the prediction error. The algorithm takes the following form: We apply

---

Algorithm 2.3 Temporal difference algorithm

Initialize $V(s) = 0$ for all $s \in S$
Initialize s arbitrarily
repeat
  For each step:
  Choose action a based on policy $\pi(s)$
  Get reward r and next state s'
  $V(s) \leftarrow V(s) + \alpha(r + \gamma V(s') - V(s))$
  $s = s'$
until Finished required number of steps or s reaches a terminal state

---

this algorithm to the previous grid game. We initialize the action probabilities to be equal in all directions; therefore, as in Section 2.7, $\pi(s, a) = 0.25$. We set the discount factor $\gamma = 0.9$, just as we did in the previous example, and set the learning rate to $\alpha = 0.001$. We then run the algorithm for 1,000,000 steps and get the value of the state as

$$V^T = \begin{bmatrix} 8.89 & 2.45 & 0.04 & 2.56 & 0.97 & -0.37 & -0.13 & -0.42 & -1.27 \end{bmatrix}.$$

Recall that the true value of the state is

$$V^T = \begin{bmatrix} 8.85 & 2.5 & -0.07 & 2.5 & 0.92 & -0.44 & -0.07 & -0.44 & -1.27 \end{bmatrix}$$

There is a major difference in what we are doing in this section and what we did in the case of value iteration. In the case of value iteration we knew the state transition probabilities $P^a_{ss'}$ and the rewards $R^a_{ss'}$, and we simply implemented an iterative search algorithm for the zero of the function but we did not actually play the game. In this section, the agent has to play the game many times to *learn* the value of the states. In the value iteration case, it required only 40 iterations to get an excellent estimate of the value function, and in this section we played the game taking $1,000,000$ steps and getting the rewards and then estimating the state value function.

2.10   TD Learning of the State-Action Function

In this section, we estimate the state-action function $Q(s, a)$. We will use essentially the same algorithm as in the previous section. However, in this case we use an ϵ-greedy action selection process. Therefore, we do not select an action based solely on the current policy, but we choose the next action based on the highest value in the state-action table. This is similar to the n-armed bandit

Table 2.1   Temporal difference Q-table learning result.

State	Actions			
	Up	Down	Right	Left
1	22.7	22.6	22.8	22.7
2	17.1	16.0	16.4	20.7
3	15.3	14.1	15.1	18.5
4	20.7	16.4	16.0	17.6
5	18.7	14.4	14.2	18.0
6	16.0	12.8	13.5	16.7
7	18.5	15.1	14.3	15.5
8	15.5	13.7	12.6	16.4
9	14.8	11.6	11.9	13.9

problem. Then with a small probability ϵ we arbitrarily choose another action; this is the exploration phase. The recursive equation is

$$Q_{k+1}(s_t, a_t) = Q_k(s_t, a_t) + \alpha(r_{t+1} + \gamma Q_k(s_{t+1}, a_{t+1}) - Q_k(s_t, a_t)) \qquad (2.46)$$

We begin the algorithm by initializing the Q-table to random numbers. In our grid game, the Q-table is a 9×4 table. In this case, the algorithm will converge to the optimal Q-table as the algorithm will adapt the action policy to the greedy action. We have set $\alpha = 0.1$ and $\gamma = 0.9$. We run this algorithm for one million steps through the grid. Table 2.1 shows the result.

Algorithm 2.4 Temporal difference state-action algorithm

1: Initialize $Q(s, a)$ arbitrarily (random numbers)
2: Initialize the state s arbitrarily
3: Choose action a based on an ϵ - greedy policy
4: **repeat**
5:     Take action a and move to the next state, s', and receive reward r.
6:     Choose the next action a' at state s', using ϵ - greedy policy
7:     $Q(s, a)) \leftarrow Q(s, a) + \alpha \left[r + \gamma Q(s', a') - Q(s, a) \right]$
8:     Set $a = a'$ and $s = s'$
9: **until** specified number of steps complete or s reaches a terminal state.

Recall, the value for the *optimal* action policy is

$$V^T = \begin{bmatrix} 24.4 & 22.0 & 19.8 & 22.0 & 19.8 & 17.8 & 19.8 & 17.8 & 16.0 \end{bmatrix}$$

State 1 has the same value regardless of the action taken because from state 1 the agent always goes to state 9 and receives a reward of +10. In this simulation, state 1 was visited nearly 200,000 times and each action was taken approximately 45,000 times and the estimated value of the state-action function is $Q(s, a) \approx 22.7$. If we look at state 2, we find that action *left* has the highest value of $Q(2, left) = 20.7$; therefore, in state 2 the best choice would be to go *left*. We can go down the rest of the table and choose the action that has the highest expected value. As such, an agent playing this grid game can look at its current state and then look up in the Q-table to find which action would give it the highest reward and then, if it were a *rational* agent, it would take the action with the highest reward.

2.11   Q-Learning

This is a well-established reinforcement learning method. It was first proposed in Reference 6 and the proof of stability of this algorithm is based on stochastic approximation and presented in Reference 7. The recursive equation for the Q-table takes the form

$$Q_{k+1}(s_t, a_t) = Q_k(s_t, a_t) + \alpha(r_{t+1} + \gamma \max_a Q_k(s_{t+1}, a) - Q_k(s_t, a_t)) \quad (2.47)$$

This algorithm is very similar to the temporal difference (TD) learning algorithm in the previous section. The key difference is that we now search for a strictly greedy policy, without exploration, when we compute $\max_a Q_k(s_{t+1}, a)$. We once again begin by initializing the state s arbitrarily and initializing the Q-table $Q(s, a)$ arbitrarily. We choose an action based on an ϵ-greedy action selection policy and then move to the next state s' and get a reward. Then, when we compute the prediction error as $r_{t+1} + \gamma \max_a Q_k(s_{t+1}, a) - Q_k(s_t, a_t)$, we do not use an ϵ-greedy algorithm, we search for the maximum value in the Q-table. The algorithm can be written as Algorithm 2.5.

Algorithm 2.5 Q-learning

Initialize $Q(s, a)$ arbitrarily (random numbers)
Initialize the state s arbitrarily
repeat
    Choose action a based on an ϵ-greedy policy
    Take action a and move to the next state, s', and receive reward r.
    $Q(s, a)) \leftarrow Q(s, a) + \alpha \left[r + \gamma \max_a Q(s', a') - Q(s, a) \right]$
    Set $s = s'$
until specified number of steps complete or s reaches a terminal state.

Table 2.2 Temporal difference Q-table learning result.

State	Actions			
	Up	Down	Right	Left
1	23.6	23.7	24.1	23.8
2	18.3	17.2	16.7	21.7
3	16.3	15.0	16.3	19.0
4	21.5	17.2	17.1	18.3
5	19.5	15.4	15.4	18.9
6	16.8	13.6	14.5	17.3
7	19.3	16.4	14.9	16.1
8	17.5	14.3	13.7	16.7
9	15.1	13.0	12.9	15.7

Once again, we set $\alpha = 0.1$ and $\gamma = 0.9$ and run this algorithm for one million steps through the grid. We get almost exactly the same Q-table as before. We then run the algorithm again, but this time we set the exploration parameter $\epsilon(k) = \epsilon(0) \div (1 + 0.000001 * k)$, where k is the step number. We then get the results illustrated in (Table 2.2). In this case, the algorithm has converged closer to the optimal values.

2.12 Eligibility Traces

In the previous sections, in the TD learning of the state-value function or the Q-table we used what is known as a *one-step predictor*. The update equations had the form

$$x_{k+1} = x_k + \alpha(r_{t+1} + \gamma y(t+1) - y(t)) \tag{2.48}$$

where the TD term $r_{t+1} + \gamma y(t+1) - y(t)$ is based only on what happened in the last time step. In other words, we only updated the effect of the last step. In the case of eligibility traces, we will look back further in time. For example, if the current reward is good, then let us not only update the current state but also give some of the reward to some of the previous states that led us to this point. This should improve the time of convergence of the algorithm. Recall, we are defining the rewards in terms of discounted future rewards as

$$R_t = r_{t+1} + \gamma r_{t+2} + \gamma^2 r_{t+3} + \cdots + \gamma^{T-t-1} = r_{t+1} + V_t(s_{t+1}) \tag{2.49}$$

A two-step predictor has the form

$$R_t = r_{t+1} + \gamma r_{t+2} + \gamma^2 V_t(s_{t+2}) \tag{2.50}$$

The eligibility trace keeps track of the last time we visited a particular state. The current reward is then assigned to recently visited states. States that have not been visited for a long time are not given much credit for the current reward.

Let us define the eligibility trace for each state at time t as $e_t(s)$. The eligibility trace for each state decays as $\gamma\alpha$, and the eligibility trace for the state just visited is increased by 1. Therefore, we update the eligibility trace as

$$e_t(s) = \begin{cases} \gamma\lambda e_{t-1}(s) & \text{if} \quad s \neq s_t \\ \gamma\lambda e_{t-1}(s) + 1 & \text{if} \quad s = s_t \end{cases} \tag{2.51}$$

The one-step prediction error is

$$\delta_t = r_{t+1} + \gamma V_t(s_{t+1}) - V_t(s_t) \tag{2.52}$$

The correction for each state becomes

$$\Delta V_t(s) = \alpha\delta_t e_t(s) \quad \forall s \tag{2.53}$$

Algorithm 2.6 TD(λ) learning

Initialize $V(s)$ arbitrarily (random numbers)
Initialize the state s arbitrarily
Initialize $e(s) = 0$
repeat
  Choose action a based on $\pi(s)$
  Take action a, observe reward and move to the next state, s'
  Compute TD error as, $\delta = r + \gamma V(s') - V(s)$
  Compute $e(s) = e(s) + 1$
  repeat
    For all states, s
    $V(s) = V(s) + \alpha\delta e(s)$
    $e(s) = \gamma\lambda e(s)$
  until All states update and set $s = s'$
until specified number of steps complete or s reaches a terminal state

---

We run the TD(λ) described in Algorithm 2.6, and set $\gamma = 0.9$, $\alpha = 0.001$ and $\lambda = 0.9$. We run through the grid for 1,000,000 steps. The estimated value of the value function is then

$$V^T = \begin{bmatrix} 8.87 & 2.49 & -0.23 & 2.5 & 0.84 & -0.49 & -0.06 & -0.47 & -1.29 \end{bmatrix}$$

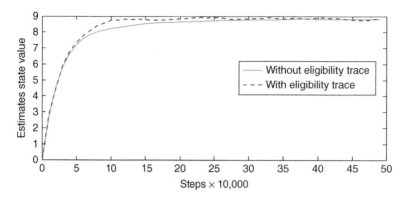

Fig. 2-7. Comparison of TD learning with and without eligibility traces for $V(1)$.

However, this is almost the same result as we got when we used the traditional TD learning algorithm described in Algorithm 2.3. The promise of using eligibility traces is the increased speed of convergence. Therefore, we run the simulation again and illustrate the convergence of $V(1)$. As illustrated in Fig. 2-7, the curve representing the eligibility trace does converge faster. Whether the improvement in the convergence rate is worth the extra effort and computational requirements is a users decision. Furthermore, in this particular example, the algorithm is not sensitive to changes in the parameter λ.

The final algorithm in this chapter is the $Q(\lambda)$ learning algorithm. In this case we apply the TD(λ) corrections to the state-action Q-table. The eligibility functions then become $e(s, a)$. Everything else is largely the same as in the previous algorithm. As usual, we will compute the prediction error as

$$\delta_t = r_{t+1} + \gamma Q_t(s_{t+1}, a_{t+1}) - Q_t(s_t, a_t) \tag{2.54}$$

and the eligibility trace becomes

$$e_t(s) = \begin{cases} \gamma \lambda e_{t-1}(s) & \text{if } s \neq s_t \\ \gamma \lambda e_{t-1}(s) + 1 & \text{if } s = s_t \end{cases} \tag{2.55}$$

and then the update for $Q(s, a)$ becomes

$$Q_{t+1}(s, a) = Q_t(s, a) + \alpha \delta_t e_t(s, a) \qquad \forall s, a \tag{2.56}$$

The algorithm is described in Algorithm 2.7.

Algorithm 2.7 $Q(\lambda)$ learning

Initialize $Q(s, a)$ arbitrarily (random numbers)
Initialize the state s arbitrarily
Initialize $e(s, a) = 0$
repeat
  For each step
  Choose action a based on $\pi(s)$
  Take action a, observe reward r and move to the next state, s'
  Choose a' from s' using an ϵ-greedy policy.
  Compute TD error as, $\delta = r + \gamma Q(s', a') - Q(s, a)$
  Compute $e(s, a) = e(s, a) + 1$
  repeat
    For all states, s and a
    $Q(s, a) = Q(s, a) + \alpha \delta e(s, a)$
    $e(s, a) = \gamma \lambda e(s, a)$
  until All states update and set $s = s'$ and $a = a'$
until specified number of steps complete or s reaches a terminal state

---

Setting $\lambda = 0.9$ and $\gamma = 0.9$ and $\epsilon = 0.1$ and $\alpha = 0.1$, we get almost the same answer as with the standard Q-learning as shown in Table 2.2. Once again, we examine the rate of convergence for the first value in the table, namely $Q(1, \text{UP})$. The graphs of the convergence of the term $Q(1, \text{UP})$ is illustrated in Fig. 2-8. The difference between the convergence rates for the Q-learning

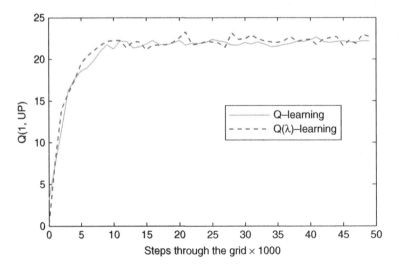

Fig. 2-8. Comparison of Q-learning with and without eligibility traces for Q(1, UP).

algorithm and the $Q(\lambda)$-learning algorithm is relatively minor in this particular case. One has to use good judgement to determine whether using the more complex $Q(\lambda)$-learning algorithm is of significant value.

References

[1] R. S. Sutton and A. G. Barto, *Reinforcement Learning: An Introduction*. Cambridge, Massachusetts: The MIT Press, 1998.

[2] M. Bowling and M. Veloso, "Multiagent learning using a variable learning rate," *Artificial Intelligence*, vol. 136, no. 2, pp. 215–250, 2002.

[3] R. Bellman, *Dynamic Programming*. Princeton, New Jersey: Princeton University Press, 1957.

[4] D. P. Bertsekas, *Dynamic Programming: Deterministic and Stochastic Models*. Englewood Cliffs, New Jersey: Prentice-Hall, 1987.

[5] R. A. Howard, *Dynamic Programming and Markov Processes*. Cambridge, Massachusetts: MIT Press, 1960.

[6] C. J. C. H. Watkins and P. Dayan, "Q-learning," *Machine Learning*, vol. 8, no. 3, pp. 279–292, 1992.

[7] T. Jaakkola, M. Jordan, and S. Singh, "On the convergence of stochasticiterative dynamic programming algorithms," *Neural Computation*, vol. 6, no. 6, pp. 1185–1201, 1994.

Chapter 3
Learning in Two-Player Matrix Games

3.1 Matrix Games

In this chapter, we will examine the two-player stage game or the matrix game problem. Now, we have two players each learning how to play the game. In some cases they may be competing with each other, or they may be cooperating with other. In this section, we will introduce the class of game that we will investigate in this chapter. In fact, almost every child has played some version of these games. We will focus on three different games: matching pennies, rock-paper-scissors, and prisoners' dilemma. These are all called *matrix games* or *stage games* because there is no state transition involved. We will limit how far we delve into game theory and focus on the learning algorithms associated with these games. The idea is for the agents to play these games repetitively and learn their best strategy. In some cases one gets a pure strategy; in other words the agent will choose the same particular action all the time, and in some cases it is best to pick an action with a particular probability, which is known as a *mixed strategy*.

In the prisoners' dilemma game, two prisoners who committed a crime together are being interrogated by the police. Each prisoner has two choices; one choice is to cooperate with the police and defect on his accomplice, and the other is to cooperate with his accomplice and lie to the police. If both

Multi-Agent Machine Learning: A Reinforcement Approach, First Edition. Howard M. Schwartz.
© 2014 John Wiley & Sons, Inc. Published 2014 by John Wiley & Sons, Inc.

Table 3.1   Examples of two-player matrix games.

(a) Matching Pennies	(b) Prisoners' Dilemma	(c) Rock-Paper-Scissors
$R_1 = \begin{bmatrix} 1 & -1 \\ -1 & 1 \end{bmatrix},$	$R_1 = \begin{bmatrix} 5 & 0 \\ 10 & 1 \end{bmatrix},$	$R_1 = \begin{bmatrix} 0 & -1 & 1 \\ 1 & 0 & -1 \\ -1 & 1 & 0 \end{bmatrix},$
$R_2 = -R_1$	$R_2 = (R_1)^{\mathrm{T}}$	$R_2 = -R_1$
NE in fully mixed strategies	NE in pure strategies	NE in fully mixed strategies

of them cooperate with each other and do not confess to the crime, then they will get just a few months in jail. If they both defect and cooperate with the police, then they will get a longer time in jail. However, if one of them defects and cooperates with the police and the other one cooperates with his accomplice and lies to the police, then the one who lied to the police and tried to cooperate with the accomplice will go to jail for a very long time. In Table 3.1, the payoff matrix for the game is shown. This matrix stipulates the rewards for player 1. In the matrix, the entries represent the rewards to the row player, and the first row represents cooperation with the accomplice and the second row represents defection and confession to the police. If the prisoners cooperate with each other and both of them pick the first row and column, then they only go to jail for a short time, a few months, and they get a good reward of 5. However, if row player defects and tells the truth to the police and the column player lies to the police and cooperates with his accomplice, the row player gets a big reward of 10 and goes free, whereas the column player would get a reward of 0 and be sent to jail for life. If they both defect and tell the truth to the police, then they each get a small reward of 1 and go to jail for a couple of years. If this was you, would you trust your criminal accomplice to cooperate with you because if he defects to the police and you lie to the police then you will go to jail for a very long time? Most rational people will confess to the police and limit the time that they may spend in jail. The choice of action to defect is known as the *Nash equilibrium* (NE). If a machine learning agent were to play this game repetitively, it should learn to play the action of *Defect* all the time, with 100% probability. This is known as a *pure* strategy game. A *pure* strategy means that one picks the same action all the time.

The next game we will define is the matching pennies game. In this game two children each hold a penny. They then independently choose to show either heads or tails. If they show two tails or two heads, then player 1 will win a reward of 1 and player 2 loses and gets a reward of −1. If they both show different sides of the coin, then player 2 wins. On any given play, one will win and one will lose. This is known as a *zero-sum matrix* game. When we say that it is a zero-sum game, we mean that one wins the same amount as the other loses. This game's *optimal* solution, or its NE, is the mixed strategy of choosing heads 50% of the time and choosing tails also 50% of the time. If player 2 always played heads, then quickly player 1 would realize that player 2 always plays heads and player 1 would also start to always play heads and would begin to win all the time. If player 2 always played heads, then we would say that player 2 was an irrational player. So clearly, each one of them should play either heads or tails 50% of the time to maximize their reward. This is known as a *mixed strategy* game; whereas in the prisoner's dilemma game the optimal strategy was always to defect 100% of the time and as such we refer to that as a *pure* strategy.

The next game of interest to us is the game of rock-paper-scissors. This game is well known to most children. The idea is to display your hand as either a rock (clenched fist), scissors, or as a flat piece of paper. Then, paper *covers* (beats) rock, rock *breaks* (beats) scissors, and scissors *cuts* (beats) paper. If both players display the same entity, then it is a tie. This game is a mixed strategy zero-sum game. The obvious solution is to randomly play each action, rock, paper, or scissors with a 33.3% probability. The only difference to this game is that we now have three actions instead of two.

More formally, a matrix game (strategic game) [1, 2] can be described as a tuple $(n, A_{1,\ldots,n}, R_{1,\ldots,n})$, where n is the agents' number, A_i is the discrete space of agent i's available actions, and R_i is the payoff function that agent i receives. In matrix games, the objective of agents is to find pure or mixed strategies that maximize their payoffs. A pure strategy is the strategy that chooses actions deterministically, whereas a mixed strategy is the strategy that chooses actions based on a probability distribution over the agent's available actions. The NE in the rock-paper-scissors game and the matching pennies game are mixed strategies that execute actions with equal probability [3].

The player i's reward function R_i is determined by all players' joint action from joint action space $A_1 \times \cdots \times A_n$. In a matrix game, each player tries to maximize its own reward based on the player's strategy. A player's strategy in a matrix game is a probability distribution over the player's action set. To evaluate a player's strategy, we introduce the following concept of NE:

Definition 3.1 *A Nash equilibrium in a matrix game is a collection of all players' strategies $(\pi_1^*, \ldots, \pi_n^*)$ such that*

$$V_i(\pi_1^*, \ldots, \pi_i^*, \ldots, \pi_n^*) \geq V_i(\pi_1^*, \ldots, \pi_i, \ldots, \pi_n^*), \quad (3.1)$$

$$\forall \pi_i \in \Pi_i, i = 1, \ldots, n \quad (3.2)$$

where $V_i(\cdot)$ is player i's value function which is player i's expected reward given all players' strategies, and π_i is any strategy of player i from the strategy space Π_i.

In other words, an NE is a collection of strategies for all players such that no player can do better by changing its own strategy given that other players continue playing their NE strategies [4]. We define $Q_i(a_1, \ldots, a_n)$ as the received reward of player i given players' joint action $a_1, \ldots, a_n$, and $\pi_i(a_i)$ $(i = 1, \ldots, n)$ as the probability of player i choosing action a_i. Then the NE defined in (3.1) becomes

$$\sum_{a_1, \ldots, a_n \in A_1 \times \cdots \times A_n} Q_i(a_1, \ldots, a_n) \pi_1^*(a_1) \cdots \pi_i^*(a_i) \cdots \pi_n^*(a_n) \geq$$

$$\sum_{a_1, \ldots, a_n \in A_1 \times \cdots \times A_n} Q_i(a_1, \ldots, a_n) \pi_1^*(a_1) \cdots \pi_i(a_i) \cdots \pi_n^*(a_n),$$

$$\forall \pi_i \in \Pi_i, i = 1, \cdots, n \quad (3.3)$$

where $\pi_i^*(a_i)$ is the probability of player i choosing action a_i under the player i's NE strategy π_i^*.

We provide the following definitions regarding matrix games:

Definition 3.2 *A Nash equilibrium is called a **strict** Nash equilibrium if (3.1) is strict [5].*

Definition 3.3 *If the probability of any action from the action set is greater than 0, then the player's strategy is called a **fully mixed strategy**.*

Definition 3.4 *If the player selects one action with probability 1 and other actions with probability 0, then the player's strategy is called a **pure strategy**.*

Definition 3.5 *A Nash equilibrium is called a **strict Nash equilibrium in pure strategies** if each player's equilibrium action is better than all its other actions, given the other players' actions [6].*

3.2 Nash Equilibria in Two-Player Matrix Games

For a two-player matrix game, we can set up a matrix with each element containing a reward for each joint action pair. Then the reward function R_i for player $i(i = 1, 2)$ becomes a matrix.

A two-player matrix game is called a *zero-sum game* if the two players are fully competitive. In this way, we have $R_1 = -R_2$. A zero-sum game has a unique NE in the sense of the expected reward. This means that, although each player may have multiple NE strategies in a zero-sum game, the value of the expected reward V_i under these NE strategies will be the same. A *general-sum matrix game* refers to all types of matrix games. In a general-sum matrix game, the NE is no longer unique and the game might have multiple NEs.

For a two-player matrix game, we define $\pi_i = (\pi_i(a_1), \ldots, \pi_i(a_{m_i}))$ as the set of all probability distributions over player i's action set $A_i(i = 1, 2)$. Then V_i becomes

$$V_i = \pi_1 R_i \pi_2^T \tag{3.4}$$

An NE for a two-player matrix game is the strategy pair (π_1^*, π_2^*) for two players such that, for $i = 1, 2$,

$$V_i(\pi_i^*, \pi_{-i}^*) \geq V_i(\pi_i, \pi_{-i}^*), \forall \pi_i \in PD(A_i) \tag{3.5}$$

where $-i$ denotes any other player than player i, and $PD(A_i)$ is the set of all probability distributions over player i's action set A_i.

Given that each player has two actions in the game, we can define a two-player two-action general-sum game as

$$R_1 = \begin{bmatrix} r_{11} & r_{12} \\ r_{21} & r_{22} \end{bmatrix}, R_2 = \begin{bmatrix} c_{11} & c_{12} \\ c_{21} & c_{22} \end{bmatrix} \tag{3.6}$$

where r_{lf} and c_{lf} denote the reward to the row player (player 1) and the reward to the column player (player 2), respectively. The row player chooses action $l \in \{1, 2\}$ and the column player chooses action $f \in \{1, 2\}$. Based on Definition 3.2 and (3.5), the pure strategies l and f are called a *strict NE in pure strategies* if

$$r_{lf} > r_{-lf}, c_{lf} > c_{l-f} \qquad \text{for } l, f \in \{1, 2\} \tag{3.7}$$

where $-l$ and $-f$ denote any row other than row l and any column other than column f, respectively.

3.3 Linear Programming in Two-Player Zero-Sum Matrix Games

One of the issues that arise in some of the machine learning algorithms is to solve for the NE. This is easier said than done. In this section, we will demonstrate how to compute the NE in competitive zero-sum games. In some of the algorithms to follow, a step in the algorithm will be to solve for the NE using linear programming or quadratic programming. To do this, we will be required to set up a constrained minimization/maximization problem that will be solved with the simplex method. The simplex method is well known in the linear programming community.

Finding the NE in a two-player zero-sum matrix game is equal to finding the minimax solution for the following equation [7]:

$$\max_{\pi_i \in PD(A_i)} \min_{a_{-i} \in A_{-i}} \sum_{a_i \in A_i} R_i \pi_i(a_i) \tag{3.8}$$

where $\pi_i(a_i)$ denotes the probability distribution over player i's action a_i, and a_{-i} denotes any action from another player other than player i. According to (3.8), each player tries to maximize the reward in the worst case scenario against its opponent. To find the solution for (3.8), one can use linear programming.

Assume we have a 2×2 zero-sum matrix game given as

$$R_1 = \begin{bmatrix} r_{11} & r_{12} \\ r_{21} & r_{22} \end{bmatrix}, R_2 = -R_1 \tag{3.9}$$

where R_1 is player 1's reward matrix and R_2 is player 2's reward matrix. We define p_j $(j = 1, 2)$ as the probability distribution over player 1's jth action and q_j as the probability distribution over player 2's jth action.

Then the linear program for player 1 is

Find (p_1, p_2) to maximize V_1

subject to

$$r_{11}p_1 + r_{21}p_2 \geq V_1 \tag{3.10}$$

$$r_{12}p_1 + r_{22}p_2 \geq V_1 \tag{3.11}$$

$$p_1 + p_2 = 1 \tag{3.12}$$

$$p_j \geq 0, \quad j = 1, 2 \tag{3.13}$$

The linear program for player 2 is

$$\text{Find } (q_1, q_2) \text{ to maximize } V_2$$

subject to

$$-r_{11}q_1 - r_{12}q_2 \geq V_2 \tag{3.14}$$

$$-r_{21}q_1 - r_{22}q_2 \geq V_2 \tag{3.15}$$

$$q_1 + q_2 = 1 \tag{3.16}$$

$$q_j \geq 0, \qquad j = 1, 2 \tag{3.17}$$

To solve the above linear programming problem, one can use the simplex method to find the optimal points geometrically. We provide three 2×2 zero-sum games below.

Example 3.1   We take the matching pennies game, for example. The reward matrix for player 1 is

$$R_1 = \begin{bmatrix} 1 & -1 \\ -1 & 1 \end{bmatrix} \tag{3.18}$$

Since $p_2 = 1 - p_1$, the linear program for player 1 becomes

$$\text{Player 1: find } p_1 \text{ to maximize } V_1$$

subject to

$$2p_1 - 1 \geq V_1 \tag{3.19}$$

$$-2p_1 + 1 \geq V_1 \tag{3.20}$$

$$0 \leq p_1 \leq 1 \tag{3.21}$$

We use the simplex method to find the solution geometrically. Figure 3-1 shows the plot of p_1 over V_1 where the gray area satisfies the constraints (3.19)–(3.21). From the plot, the maximum value of V_1 within the gray area is 0 when $p_1 = 0.5$. Therefore, $p_1 = 0.5$ is the Nash equilibrium strategy for player 1. Similarly, we can use the simplex method to find the Nash equilibrium strategy for player 2. After solving (3.14)–(3.17), we can find that the maximum value of V_2 is 0 when $q_1 = 0.5$. Then this game has a Nash equilibrium ($p_1 = 0.5$, $q_1 = 0.5$), which is a fully mixed strategy Nash equilibrium.

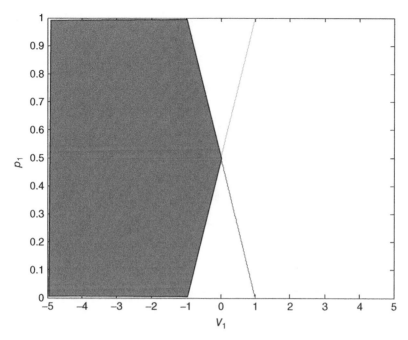

Fig. 3-1. Simplex method for player 1 in the matching pennies game. Reproduced from [8], © X. Lu.

Example 3.2   We change the reward r_{12} from -1 in (3.18) to 2 and call this game as the revised version of the matching pennies game. The reward matrix for player 1 becomes

$$R_1 = \begin{bmatrix} 1 & 2 \\ -1 & 1 \end{bmatrix} \tag{3.22}$$

The linear program for player 1 is

Player 1: find p_1 to maximize V_1

subject to

$$2p_1 - 1 \geq V_1 \tag{3.23}$$

$$p_1 + 1 \geq V_1 \tag{3.24}$$

$$0 \leq p_1 \leq 1 \tag{3.25}$$

From the plot in Fig. 3-2, we can find that the maximum value of V_1 in the gray area is 1 when $p_1 = 1$. Similarly, we can find the maximum value of

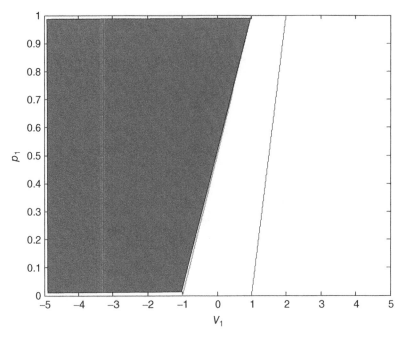

Fig. 3-2. Simplex method for player 1 in the revised matching pennies game. Reproduced from [8], © X. Lu.

$V_2 = -1$ when $q_1 = 1$. Therefore, this game has a Nash equilibrium ($p_1 = 1$, $q_1 = 1$), which is a pure strategy Nash equilibrium.

Example 3.3   We now consider the following zero-sum matrix game:

$$R_1 = \begin{bmatrix} r_{11} & 2 \\ 3 & -1 \end{bmatrix}, R_2 = -R_1 \qquad (3.26)$$

where $r_{11} \in \mathfrak{R}$. Based on different values of r_{11}, we want to find the Nash equilibrium strategies (p_1, q_1). The linear program for each player becomes

Player 1: Find p_1 to maximize V_1

subject to

$$(r_{11} - 3)p_1 + 3 \geq V_1 \qquad (3.27)$$

$$3p_1 - 1 \geq V_1 \qquad (3.28)$$

$$0 \leq p_1 \leq 1 \qquad (3.29)$$

Player 2: Find q_1 to maximize V_2

subject to

$$(2 - r_{11})q_1 - 2 \geq V_2 \tag{3.30}$$

$$-4q_1 + 1 \geq V_2 \tag{3.31}$$

$$0 \leq q_1 \leq 1 \tag{3.32}$$

We use the simplex method to find the Nash equilibria for the players with a varying r_{11}. When $r_{11} > 2$, we find that the Nash equilibrium is in pure strategies ($p_1^* = 1, q_1^* = 0$). When $r_{11} < 2$, we find that the Nash equilibrium is in fully mixed strategies ($p_1^* = 4/(6 - r_{11}), q_1^* = 3/(6 - r_{11})$). For $r_{11} = 2$, we plot the players' strategies over their value functions in Fig. 3-3. From the plot we find that player 1's Nash equilibrium strategy is $p_1 = 1$, and player 2's Nash equilibrium strategy is $q_1 \in [0, 0.75]$, which is a set of strategies. Therefore, at $r_{11} = 2$, we have multiple Nash equilibria which are $p_1 = 1, q_1 \in [0, 0.75]$. We also plot the Nash equilibria (p_1, q_1) over r_{11} in Fig. 3-4.

3.4   The Learning Algorithms

In this section, we will present several algorithms that have gained popularity within the field of machine learning. We will focus on the algorithms that have been used for learning how to choose the *optimal* actions when agents are playing matrix games. Once again, these algorithms will look like gradient descent (ascent) algorithms. We will discuss their strengths and weaknesses. In particular, we are going to look at the gradient ascent (GA) algorithm and its related version the infinitesimal gradient ascent (IGA) algorithm and the policy hill climbing (PHC) algorithm and the variable learning rate version called the *win or learn fast-policy hill climbing* (WoLF-PHC) algorithm [3]. We will then examine the linear reward-inaction (L_{RI}) and the lagging anchor algorithm. Finally, we will discuss the advantages of the L_{RI} lagging anchor algorithm. There are a number of versions of these algorithms in the literature, but they tend to be minor variations of the ones being discussed here. Of course, one could argue that all learning algorithms are minor variations of the stochastic approximation technique.

3.5   Gradient Ascent Algorithm

One of the fundamental algorithms associated with learning in matrix games is the GA algorithm and its related formulation called the *IGA* algorithm. This algorithm is used in relatively simple two-action/two-player general-sum

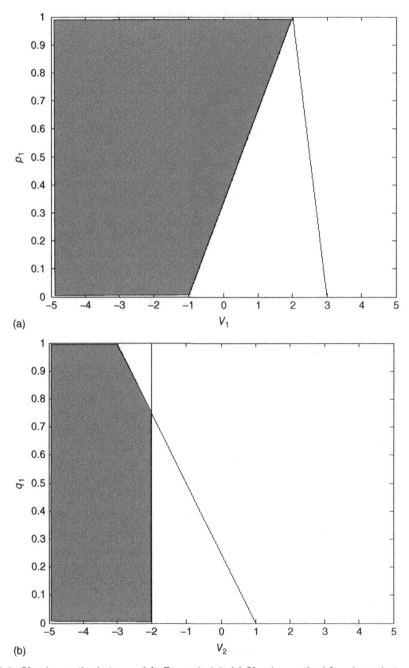

(a)

(b)

Fig. 3-3. Simplex method at $r_{11} = 2$ in Example 3.3. (a) Simplex method for player 1 at $r_{11} = 2$. (b) Simplex method for player 2 at $r_{11} = 2$. Reproduced from [8], © X. Lu.

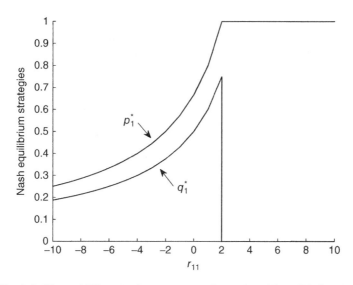

Fig. 3-4. Players' NE strategies versus r_{11}. Reproduced from [8], © X. Lu.

games. Theoretically, this algorithm will fail to converge. It can be shown that by introducing a variable learning rate that tends to zero as $\lim_{t\to\infty}\eta \to 0$, the GA algorithm will converge.

We will examine the GA algorithm presented by Singh et al. [9]. We examine the case of a 2×2 matrix game as two payoff matrices, one for the row player and one for the column player. The matrices are

$$R_r = \begin{bmatrix} r_{11} & r_{12} \\ r_{21} & r_{22} \end{bmatrix} \tag{3.33}$$

and

$$R_c = \begin{bmatrix} c_{11} & c_{12} \\ c_{21} & c_{22} \end{bmatrix} \tag{3.34}$$

Then, if the row player chooses action 1 and the column player chooses action 2, then the reward to player 1 (the row player) is r_{12} and the reward to player 2 (the column player) is c_{12}. This is a two-action two-player game and we are assuming the existence of a mixed strategy, although the algorithm can be used for pure strategy games as well. In a mixed strategy game, the probability that the row player chooses action 1 is $P\{a_r = 1\} = \alpha$ and, therefore, the probability that the row player chooses action 2 must be given by $P\{a_r = 2\} = 1 - \alpha$. Similarly, for player 2 (the column player), the probability that player 2 chooses action 1 is given by $P\{a_c = 1\} = \beta$ and, therefore, the probability of choosing action 2 is $P\{a_c = 1\} = 1 - \beta$. The strategy of the matrix game is completely defined by the joint strategy $\pi(\alpha, \beta)$, where α and β are constrained to remain within the unit square. We define the expected payoff to each player as $V_r(\alpha, \beta)$ and $V_c(\alpha, \beta)$. We can write the expected payoffs as

$$V_r(\alpha, \beta) = \alpha\beta r_{11} + \alpha(1 - \beta)r_{12} + (1 - \alpha)\beta r_{21} \tag{3.35}$$

$$+(1 - \alpha)(1 - \beta)r_{22} \tag{3.36}$$

$$= u_r\alpha\beta + \alpha(r_{12} - r_{22}) + \beta(r_{21} - r_{22}) + r_{22} \tag{3.37}$$

$$V_c(\alpha, \beta) = \alpha\beta c_{11} + \alpha(1 - \beta)c_{12} + (1 - \alpha)\beta c_{21} \tag{3.38}$$

$$+(1 - \alpha)(1 - \beta)c_{22} \tag{3.39}$$

$$= u_c\alpha\beta + \alpha(c_{12} - c_{22}) + \beta(c_{21} - c_{22}) + c_{22} \tag{3.40}$$

where

$$u_r = r_{11} - r_{12} - r_{21} + r_{22} \tag{3.41}$$

$$u_c = c_{11} - c_{12} - c_{21} + c_{22} \tag{3.42}$$

We can now compute the gradient of the payoff function with respect to the strategy as

$$\frac{\partial V_r(\alpha, \beta)}{\partial \alpha} = \beta u_r + (r_{12} - r_{22}) \tag{3.43}$$

$$\frac{\partial V_c(\alpha, \beta)}{\partial \beta} = \alpha u_c + (c_{21} - c_{22}) \tag{3.44}$$

The GA algorithm then becomes

$$\alpha_{k+1} = \alpha_k + \eta \frac{\partial V_r(\alpha_k, \beta_k)}{\partial \alpha_k} \tag{3.45}$$

$$\beta_{k+1} = \beta_k + \eta \frac{\partial V_c(\alpha_k, \beta_k)}{\partial \beta_k} \tag{3.46}$$

Theorem 3.1   If both players follow infinitesmal gradient ascent (IGA), where $\eta \to 0$, then their strategies will converge to a Nash equilibrium, or the average payoffs over time will converge in the limit to the expected payoffs of a Nash equilibrium.

The first algorithm we will try is the GA algorithm. We will play the mixed strategy games of matching pennies. To implement the GA learning algorithm for the matching pennies game, one needs to know the payoff matrix in

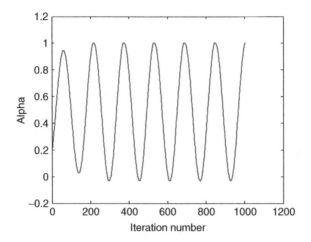

Fig. 3-5. GA in matching pennies game.

advance. One can see from Fig. 3-5 that the strategy oscillates between 0 and 1. If we try to implement the IGA algorithm, one runs into the difficulty of trying to choose an appropriate rate of convergence of the step size to zero. This is not a practical algorithm to use. Therefore, the GA algorithm does not work particularly well; it oscillates and one can show this theoretically [3].

3.6 WoLF-IGA Algorithm

The WoLF-IGA algorithm was introduced by Bowling and Veloso [3] for two-player two-action matrix games. As a GA learning algorithm, the WoLF-IGA algorithm allows the player to update its strategy based on the current gradient and a variable learning rate. The value of the learning rate is smaller when the player is winning, and it is larger when the player is losing. The term p_1 is the probability of player 1 choosing the first action. Then, $1 - p_1$ is the probability of player 1 choosing the second action. Accordingly, q_1 is the probability of player 2 choosing the first action and $1 - q_1$ is the probability of player 2 choosing the second action. The updating rules of the WoLF-IGA algorithm are as follows:

$$p_1(k + 1) = p_1(k) + \eta\alpha_1(k)\frac{\partial V_1(p_1(k), q_1(k))}{\partial p_1} \tag{3.47}$$

$$q_1(k + 1) = q_1(k) + \eta\alpha_2(k)\frac{\partial V_2(p_1(k), q_1(k))}{\partial q_1} \tag{3.48}$$

$$\alpha_1(k) = \begin{cases} \alpha_{\min}, & \text{if } V_1(p_1(k), q_1(k)) > V_1(p_1^*, q_1(k)) \\ \alpha_{\max}, & \text{otherwise} \end{cases}$$

$$\alpha_2(k) = \begin{cases} \alpha_{\min}, & \text{if } V_2(p_1(k), q_1(k)) > V_2(p_1(k), q_1^*) \\ \alpha_{\max}, & \text{otherwise} \end{cases}$$

where η is the step size, $\alpha_i(i = 1, 2)$ is the learning rate for player $i(i = 1, 2)$, $V_i(p_1(k), q_1(k))$ is the expected reward of player i at time k given the current two players' strategy pair $(p_1(k), q_1(k))$, and (p_1^*, q_1^*) are equilibrium strategies for the players. In a two-player two-action matrix game, if each player uses the WoLF-IGA algorithm with $\alpha_{\max} > \alpha_{\min}$, the players' strategies converge to an NE as the step size $\eta \to 0$ [3].

This algorithm is a GA learning algorithm that can guarantee the convergence to an NE in fully mixed or pure strategies for two-player two-action general-sum matrix games. However, this algorithm is not a decentralized learning algorithm. It requires the knowledge of $V_1(p_1^*, q_1(k))$ and $V_2(p_1(k), q_1^*)$ in order to choose the learning parameters $\alpha_{\min}$ and $\alpha_{\max}$ accordingly. In order to obtain $V_1(p_1^*, q_1(k))$ and $V_2(p_1(k), q_1^*)$, we need to know each player's reward matrix and its opponent's strategy at time k; whereas in a decentralized learning algorithm, the agents would only have their own actions and reward at time k. Although a practical decentralized learning algorithm called a *WoLF-PHC method* was provided in Reference 3, there is no proof of convergence to NE strategies.

3.7   Policy Hill Climbing (PHC)

A more practical version of the gradient descent algorithm is the PHC algorithm. This algorithm is based on the Q-learning algorithm that we presented in Chapter 2. This is a rational algorithm that can estimate mixed strategies. The algorithm will converge to the *optimal* mixed strategies if the other players are not learning and are therefore playing *stationary* strategies.

The PHC algorithm is a simple practical algorithm that can learn mixed strategies. Hill climbing is performed by the PHC algorithm in the space of the mixed strategies. This algorithm was first proposed by Bowling and Veloso [3]. The PHC does not require much information as neither the recent actions executed by the agent nor the current strategy of its opponent is required to be known. The probability that the agent selects the highest valued actions is increased by a small learning rate $\delta \in (0,1]$. The algorithm is equivalent to the single-agent Q-learning when $\delta = 1$ as the policy moves to the greedy policy with probability 1. The PHC algorithm is rational and converges to the optimal solution when a fixed (stationary) strategy is followed by the other players. However, the PHC algorithm may not converge to a stationary policy if the other players are learning [3].

The convergence proof is the same as for Q-learning [10], which guarantees that the Q values will converge to the optimal Q^* with a suitable exploration policy [9]. However, when both players are learning, then the algorithm will not necessarily converge. The algorithm starts from the Q-learning algorithm and is given as

$$Q_{t+1}^j(a) = (1 - \alpha)Q_t^j(a) + \alpha(r^j + \gamma \max_{a'} Q_t^j(a')) \tag{3.49}$$

$$\pi_{t+1}^j(a) = \pi_t^j(a) + \Delta_a \tag{3.50}$$

where

$$\Delta_a = \begin{cases} -\delta_a & \text{if } a \neq \text{argmax}_{a'} Q_t^j(a') \\ \sum_{a' \neq a} \delta_{a'} & \text{otherwise} \end{cases}$$

where $\delta_a = \min\left(\pi_t^j(a), \frac{\delta}{|A_j|-1}\right)$

The algorithm is given as,

Algorithm 3.1 Policy hill-climbing (PHC) algorithm for agent j:

Initialize:
learning rates $\alpha \in (0,1]$, $\delta \in (0,1]$
discount factor $\gamma \in (0,1)$
exploration rate ϵ
$Q^j(a) \leftarrow 0 \quad$ and $\quad \pi^j(a) \leftarrow \frac{1}{|A_j|}$
Repeat
(a) Select an action a according to the strategy $\pi_t^j(a)$ with some exploration ϵ.
(b) Observe the immediate reward r^j.
(c) Update $Q_{t+1}^j(a)$ using Eq. (3.49).
(d) Update the strategy $\pi_{t+1}^j(a)$ by using Eq. (3.50).

We will now run a simulation of the matching pennies games. To generate the simulation results illustrated in Fig. 3-6, we set the learning rate $\alpha = 1/(10 + 0.00001t)$, the exploration rate to $\epsilon = 0.5/(1 + 0.0001t)$, and $\delta = 0.0001$. We initialize the probability of player 1 choosing action 1, at 80%. One can see that the algorithm will oscillate about the NE as expected by the theory. In this case, both players are learning. For any practical application, this is a poor

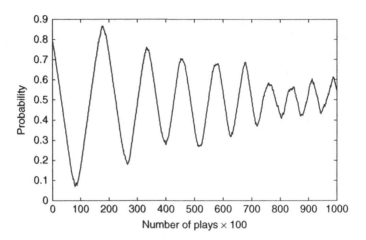

Fig. 3-6. PHC matching pennies game, player 1, probability of choosing action 1, heads.

result. Furthermore, it takes many iterations to converge about the 50% equilibrium point. Another issue with implementing this algorithm is choosing all the parameters. In a more complex game, this algorithm would not be practical to implement.

In the next case, we set the column player to always play heads, action 1, and we start the row player at 20% heads and 80% tails. Then the row player should learn to always play heads 100% of the time. As illustrated in Fig. 3-7, the probability of player 1 choosing heads increases and converges to a probability of 100%.

3.8   WoLF-PHC Algorithm

In Reference 3, the authors propose to use a variable learning rule as

$$\alpha_{k+1} = \alpha_k + \eta l_k^r \frac{\partial V_r(\alpha_k, \beta_k)}{\partial \alpha_k} \qquad (3.51)$$

$$\beta_{k+1} = \beta_k + \eta l_k^c \frac{\partial V_c(\alpha_k, \beta_k)}{\partial \beta_k} \qquad (3.52)$$

where the term l is a variable learning rate given by $l \in [l_{\min}, l_{\max}] > 0$.

The method for adjusting the learning rate l is referred to as the *WoLF* approach. The idea is when one is winning the game to adjust the learning rate to learn slowly and be cautious, and when losing or doing poorly to learn quickly. The next step is to determine when the agent is doing well or doing

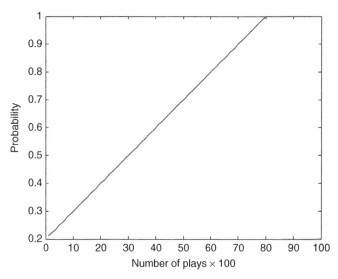

Fig. 3-7. PHC matching pennies game, player 1, probability of choosing action 1, heads when player 2 always chooses heads.

poorly in playing the game. The conceptual idea is for the agent to choose an NE and compare the expected reward it would receive to the NE. If the reward it would receive is greater than the NE, then it is winning and will learn slowly and cautiously. Otherwise, it is losing and it should learn fast; the agent does want to be losing.

The two players each select an NE of their choice independently; they do not need to choose the same equilibrium point. If there are multiple NE points in the game, then the agents could pick different points; that is perfectly acceptable because each NE point will have the same value. Therefore, player 1 may choose NE point α^e and player 2 may choose NE point β^e, and the learning rates are chosen as

$$l_k^r = \begin{cases} l_{\min} & \text{if } V_r(\alpha_k, \beta_k) > V_r(\alpha^e, \beta_k) & \text{Winning} \\ l_{\max} & \text{otherwise} & \text{Losing} \end{cases}$$

$$l_k^c = \begin{cases} l_{\min} & \text{if } V_c(\alpha_k, \beta_k) > V_c(\alpha_k, \beta^e) & \text{Winning} \\ l_{\max} & \text{otherwise} & \text{losing} \end{cases}$$

When we combine the variable learning rate with the IGA algorithm, we refer to it as the *WoLF-IGA algorithm*. Although this is not a practical algorithm to implement, it does have good theoretical properties as defined by the following theorem.

Theorem 3.2   If in a two-action iterated general-sum game both players follow the WoLF-IGA algorithm (with $l_{max} > l_{min}$), then their strategies will converge to a Nash Equilibrium.

It is interesting to note that winning is defined as the expected reward of the current strategy being greater than the expected reward of the current player's NE strategy and the other player's current strategy.

The difficulty with the WoLF-IGA algorithm is the amount of information that the player must have. The player needs to know its own payoff matrix, the other player's strategy, and its own NE. Of course, if one knows its own payoff matrix, then it will also know its NE point or points. That is a lot of information for the player to know, and as such this is not a practical algorithm to implement.

The WoLF-PHC algorithm is an extension of the PHC algorithm [3]. This algorithm uses the mechanism of win-or-learn-fast (WoLF) so that the PHC algorithm converges to an NE in self-play. The algorithm has two different learning rates, δ_w when the algorithm is winning and δ_l when it is losing. The difference between the average strategy and the current strategy is used as a criterion to decide when the algorithm wins or loses. The learning rate δ_l is larger than the learning rate δ_w. As such, when a player is losing, it learns faster than when winning. This causes the player to adapt quickly to the changes in the strategies of the other player when it is doing more poorly than expected and learns cautiously when it is doing better than expected. This also gives the other player the time to adapt to the player's strategy changes. The WoLF-PHC algorithm exhibits the property of convergence as it makes the player converge to one of its NEs. This algorithm is also a rational learning algorithm because it makes the player converge to its optimal strategy when its opponent plays a stationary strategy. These properties permit the WoLF-PHC algorithm to be widely applied to a variety of stochastic games [3, 11–13]. The recursive Q-learning of a learning agent j is given as

$$Q_{t+1}^j(a) = (1 - \alpha)Q_t^j(a) + \alpha(r^j + \gamma \max_{a'} Q_t^j(a')) \qquad (3.53)$$

The WoLF-PHC algorithm updates the strategy of the agent j by equation 3.54, whereas Algorithm 2.1 describes the complete formal definition of the WoLF-PHC algorithm for a learning agent j:

$$\pi_{t+1}^j(a) = \pi_t^j(a) + \Delta_a \qquad (3.54)$$

where

$$
\Delta_a = \begin{cases} -\delta_a & \text{if } a \neq \text{argmax}_{a'} Q_t^j(a') \\ \sum_{a' \neq a} \delta_{a'} & \text{otherwise} \end{cases} \qquad \delta_a = \min\left(\pi_t^j(a), \frac{\delta}{|A_j| - 1}\right)
$$

$$
\delta = \begin{cases} \delta_w & \text{if } \sum_{a'} \pi_t(a') Q_{t+1}^j(a') > \\ & \sum_{a'} \bar{\pi}_{t+1}(a') Q_{t+1}^j(a') \\ \delta_l & \text{otherwise} \end{cases}
$$

$$
\bar{\pi}_{t+1}^j(a') = \bar{\pi}_t^j(a') + \frac{1}{C_{t+1}}(\pi_t^j(a') - \bar{\pi}_t^j(a')) \quad \forall a' \in A_j
$$

$$
C_{t+1} = C_t + 1
$$

Algorithm 3.2 The win-or-learn-fast policy hill climbing (WoLF-PHC) algorithm for agent j

---

Initialize:
learning rates $\alpha \in (0,1]$, $\delta_w \in (0,1]$ and $\delta_l > \delta_w$
discount factor $\gamma \in (0,1)$
exploration rate ϵ
$Q^j(a) \leftarrow 0 \quad$ and $\quad \pi^j(a) \leftarrow \frac{1}{|A_j|}$
$C(s) \leftarrow 0$
Repeat
(a) Select an action a according to the strategy $\pi_t^j(a)$ with some exploration ϵ.
(b) Observe the immediate reward r^j.
(c) Update $Q_{t+1}^j(a)$ using Eq. (3.53).
(d) Update the strategy $\pi_{t+1}^j(a)$ by using Eq. (3.54).

We simulate the WoLF-PHC algorithm for the matching pennies game. We set the learning parameter $\alpha = 1/(10 + 0.00001t)$ and $\delta_w = 1/(20000 + j)$ and $\delta_l = 2\delta_w$. The strategy for player 1 is initially set to $\pi_r = [0.2 \ 0.8]$ and the strategy for player 2 to $\pi = [0.5 \ 0.5]$. The results are shown in Fig. 3-8.

3.9   Decentralized Learning in Matrix Games

Decentralized learning means that there is no central learning strategy for all of the agents. Instead, each agent learns its own strategy. Decentralized learning

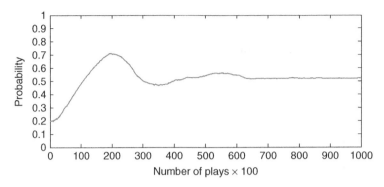

Fig. 3-8. WoLF-PHC matching pennies game, player 1, probability of choosing action 1.

algorithms can be used by players to learn their NEs in games with incomplete information [14, 15]. When an agent has "incomplete information," it means that the agent knows neither its own reward function, nor the other players' strategies, nor the other players' reward functions. The agent only knows its own action and the received reward at each time step. The main challenge for designing a decentralized learning algorithm with incomplete information is to prove that the players' strategies converge to an NE.

There are a number of multiagent learning algorithms proposed in the literature that can be used for two-player matrix games. Lakshmivarahan and Narendra [14] presented a linear reward–inaction approach that can guarantee the convergence to a NE under the assumption that the game only has strict NEs in pure strategies. The linear reward–penalty approach, introduced in Reference 15, can guarantee that the expected value of players' strategies converges to an NE in fully mixed strategies with the proper choice of parameters. Bowling and Veloso proposed a WoLF-IGA approach that can guarantee the convergence to an NE for two-player two-action matrix games and the NE can be in fully mixed strategies or in pure strategies. However, the WoLF-IGA approach is not a completely decentralized learning algorithm because the player has to know its opponent's strategy at each time step. Dahl [16, 17] proposed a lagging anchor model approach that can guarantee the convergence to an NE in fully mixed strategies. But the lagging anchor algorithm is not a decentralized learning algorithm because each player has to know its reward matrix.

We evaluate the learning automata algorithm L_{R-I} [14] and L_{R-P} [15], the GA algorithm WoLF-IGA [3], and the lagging anchor algorithm [16]. We then propose the new L_{R-I} lagging anchor algorithm. The L_{R-I} lagging anchor algorithm is a combination of learning automata and GA learning. It is a completely decentralized algorithm and, therefore, each agent only needs to know its own action and its own reward at each time step. We prove the convergence of the L_{R-I} lagging anchor algorithm to NEs in two-player two-action general-sum matrix games. Furthermore, the NE can be in games with pure or fully mixed strategies. We then simulate three matrix games to test the performance of the L_{R-I} lagging anchor learning algorithm.

We first review the multiagent learning algorithms in matrix games based on the learning automata scheme and the GA schemes. In Section 3.14, we introduce the new L_{R-I} lagging anchor algorithm and provide the proof of convergence to NEs in two-player two-action general-sum matrix games. Simulations of three matrix games are also illustrated in Section 3.14 to show the convergence of our proposed L_{R-I} lagging anchor algorithm.

3.10   Learning Automata

Learning in a two-player matrix game can be expressed as the process of each player updating its strategy according to the received reward from the environment. A learning scheme is used for each player to update its own strategy toward a NE based on the information from the environment. In order to address the limitations of the previously published multiagent learning algorithms for matrix games, we divide these learning algorithms into two groups. One group is based on learning automata [18], and another group is based on GA learning [9].

Learning automation is a learning unit for adaptive decision making in an unknown environment [18]. The objective of learning automation is to learn the optimal action or strategy by updating its action probability distribution based on the environment response. The learning automata approach is a completely decentralized learning algorithm because each learner only considers its action and the received reward from the environment and ignores any information from other agents such as the actions taken by other agents. The learning automation can be represented as a tuple (A, r, p, U), where $A = \{a_1, \ldots, a_m\}$ is the player's action set, $r \in [0, 1]$ is the reinforcement signal, p is the probability distribution over the actions, and U is the learning algorithm to update p. There are two typical learning algorithms based on learning automata: the linear reward–inaction (L_{R-I}) algorithm and the linear reward–penalty (L_{R-P}) algorithm.

3.11   Linear Reward–Inaction Algorithm

The linear reward–inaction (L_{R-I}) algorithm for player $i(i = 1, \ldots, n)$ is defined as follows:

$$p_c^i(k+1) = p_c^i(k) + \eta r^i(k)(1 - p_c^i(k)) \quad \text{if } a_c \text{ is the current action at } k$$

$$p_j^i(k+1) = p_j^i(k) - \eta r^i(k)p_j^i(k) \qquad \text{for all } a_j^i \neq a_c^i \qquad (3.55)$$

where k is the time step, the superscripts and subscripts on p denote different players and each player's different action, respectively, $0 < \eta < 1$ is the learning parameter, $r^i(k)$ is the response of the environment given player i's action a_c^i at k, and p_c^i is the probability distribution over player i's action $a_c^i(c = 1, \ldots, m)$.

In a matrix game with n players, if each player uses the L_{R-I} algorithm, then the L_{R-I} algorithm guarantees the convergence to a NE under the assumption that the game only has strict NEs in pure strategies [14].

3.12   Linear Reward–Penalty Algorithm

The linear reward–penalty (L_{R-P}) algorithm for player i is defined as follows:

$$p_c^i(k+1) = p_c^i(k) + \eta_1 r^i(k)[1 - p_c^i(k)] - \eta_2[1 - r^i(k)]p_c^i(k)$$

$$p_j^i(k+1) = p_j^i(k) - \eta_1 r^i(k)p_j^i(k) + \eta_2[1 - r^i(k)]\left[\frac{1}{m-1} - p_j^i(k)\right] \text{(for all } a_j^i \neq a_c^i)$$

$$(3.56)$$

where a_c^i is the current action the player i has taken, $0 < \eta_1, \eta_2 < 1$ are learning parameters, and m is the number of actions in the player's action set.

In a two-player zero-sum matrix game, if each player uses the L_{R-P} and chooses $\eta_2 < \eta_1$, then the expected value of the fully mixed strategies for both players can be made arbitrarily close to an NE [15]. This means that the L_{R-P} algorithm can guarantee the convergence to an NE in the sense of expected value but not the player's strategy itself.

3.13   The Lagging Anchor Algorithm

The lagging anchor algorithm for two-player zero-sum games was introduced by Dahl [16]. As a GA learning method, the lagging anchor algorithm updates the players' strategies according to the gradient. We denote player 1's strategy as a vector $\mathbf{v} = [p_1, p_2, \ldots, p_{m_1}]^T$, which is the probability distribution over all the possible actions. Accordingly, player 2's strategy is denoted as a vector $\mathbf{w} = [q_1, q_2, \ldots, q_{m_2}]^T$. The updating rules are listed as follows:

$$\mathbf{v}(k+1) = \mathbf{v}(k) + \eta \mathbf{P}_{m_1} R_1 Y(k) + \eta\gamma(\bar{\mathbf{v}}(k) - \mathbf{v}(k))$$

$$\bar{\mathbf{v}}(k) = \bar{\mathbf{v}}(k) + \eta\gamma(\mathbf{v}(k) - \bar{\mathbf{v}}(k))$$

$$\mathbf{w}(k+1) = \mathbf{w}(k) + \eta \mathbf{P}_{m_2} R_2 X(k) + \eta\gamma(\bar{\mathbf{w}}(k) - \mathbf{w}(k))$$

$$\bar{\mathbf{w}}(k) = \bar{\mathbf{w}}(k) + \eta\gamma(\mathbf{w}(k) - \bar{\mathbf{w}}(k)) \qquad (3.57)$$

where η is the step size, $\gamma > 0$ is the anchor drawing factor, and $\mathbf{P}_{m_i} = \mathbf{I}_{m_i} - (1/m_i)\mathbf{1}_{m_i}\mathbf{1}_{m_i}^T$ is a matrix used to maintain the summation of the elements in the vector $\mathbf{v}$ or $\mathbf{w}$ to be 1. $Y(k)$ is a unit vector corresponding to the actions of player 2. If the m_ith action in player 2's action set is selected at time k, then the m_ith element in $Y(k)$ is set to 1 and the other elements in $Y(k)$ are zeros. Similarly, $X(k)$ is the unit vector corresponding to the actions of player 1, and R_1 and R_2 are the reward matrices for player 1 and 2, respectively. In (3.57), $\bar{\mathbf{v}}$ and $\bar{\mathbf{w}}$ are the anchor parameters for $\mathbf{v}$ and $\mathbf{w}$, respectively, which can be represented as the weighted average of the players' strategies. In a two-player zero-sum game with only NEs in fully mixed strategies, if each player uses the lagging anchor algorithm, then the players' strategies converge to an NE as the step size $\eta \to 0$ [17].

This algorithm guarantees the convergence to an NE in fully mixed strategies. However, the convergence to an NE in pure strategies has never been discussed. Furthermore, the lagging anchor algorithm in (3.57) requires full information of the player's reward matrices R_1 and R_2. Therefore, the lagging anchor algorithm is not a decentralized learning algorithm.

Table 3.2 compares these algorithms based on the allowable number of actions for each player, the convergence to pure strategies or fully mixed strategies, and the level of decentralization. From this table, only the WoLF-IGA algorithm can guarantee the convergence to both pure and mixed-strategy NE. But it is not a decentralized learning algorithm. Although the L_{R-I} algorithm and the L_{R-P} algorithm are decentralized learning algorithms, neither of them can guarantee the convergence to both pure and mixed-strategy NE. The L_{R-I} lagging anchor algorithm presented in the next section can guarantee convergence of both pure and mixed strategies to the NE as shown in Table 3.2.

Table 3.2 Comparison of learning algorithms in matrix games.

	Existing algorithms				Our proposed algorithm
Applicability	L_{R-I}	L_{R-P}	WoLF-IGA	Lagging anchor	L_{R-I} lagging anchor
Allowable actions	No limit	Two actions	Two actions	No limit	Two actions
Convergence	Pure NE	Fully mixed NE (expected value)	Both	Fully mixed NE	Both
Decentralized?	Yes	Yes	No	No	Yes

3.14 L_{R-I} Lagging Anchor Algorithm

In this section, we design an L_{R-I} lagging anchor algorithm which is a completely decentralized learning algorithm and can guarantee the convergence to NEs in both pure and fully mixed strategies. We take the L_{R-I} algorithm defined in (3.55) as the updating law of the player's strategy and add the lagging anchor term in (3.57). Then the L_{R-I} lagging anchor algorithm for player i is defined as follows:

$$\left.\begin{aligned} p_c^i(k+1) &= p_c^i(k) + \eta r^i(k)[1 - p_c^i(k)] + \eta[\bar{p}_c^i(k) - p_c^i(k)] \\ \bar{p}_c^i(k+1) &= \bar{p}_c^i(k) + \eta[p_c^i(k) - \bar{p}_c^i(k)] \end{aligned}\right\} \begin{aligned} &\text{if } a_c^i \text{ is the action} \\ &\text{taken at time } k \end{aligned}$$

$$\left.\begin{aligned} p_j^i(k+1) &= p_j^i(k) - \eta r^i(k)p_j^i(k) + \eta[\bar{p}_j^i(k) - p_j^i(k)] \\ \bar{p}_j^i(k+1) &= \bar{p}_j^i(k) + \eta[p_j^i(k) - \bar{p}_j^i(k)] \end{aligned}\right\} \text{for all } a_j^i \neq a_c^i \quad (3.58)$$

where η is the step size and $(\bar{p}_c^i, \bar{p}_j^i)$ are the lagging parameters for (p_c^i, p_j^i). The idea behind the L_{R-I} lagging anchor algorithm is that we consider both the player's current strategy and the long-term average of the player's previous strategies at the same time. We expect that the player's current strategy and the long-term average will be drawn toward the equilibrium point during learning.

To analyze the above L_{R-I} lagging anchor algorithm, we use ordinary differential equations (ODEs). The behavior of the learning algorithm can be approximated by ODEs as the step size goes to zero. Thathachar and Sastry [19] provided the equivalent ODEs of the L_{R-I} algorithm in (3.55) as

$$\dot{p}_c^i = \sum_{j=1}^{m_i} p_c^i p_j^i (d_c^i - d_j^i) \quad (3.59)$$

where d_c^i is the expected reward given that player i is choosing action a_c^i and the other players are following their current strategies.

Combining the above ODEs of the L_{R-I} algorithm in (3.59) with the ODEs for the lagging anchor part of our algorithm, we can find the equivalent ODEs for our L_{R-I} lagging anchor algorithm, given as

$$\dot{p}_c^i = \sum_{j=1}^{m_i} p_c^i p_j^i (d_c^i - d_j^i) + (\bar{p}_c^i - p_c^i)$$

$$\dot{\bar{p}}_c^i = p_c^i - \bar{p}_c^i \quad (3.60)$$

Based on our proposed L_{R-I} lagging anchor algorithm, we now present the following theorem:

Theorem 3.3 We consider a two-player two-action general-sum matrix game and assume the game only has a Nash equilibrium in fully mixed strategies or strict Nash equilibria in pure strategies. If both players follow the L_{R-I} lagging anchor algorithm, when the step size $\eta \to 0$, then the following is true regarding the asymptotic behavior of the algorithm:

- All Nash equilibria are asymptotically stable;
- Any equilibrium point which is not a Nash equilibrium is unstable.

Proof: Given a two-player two-action general-sum game defined in (3.6), we denote p_1 as the probability of player 1 taking its first action and q_1 as the probability of player 2 taking its first action. Then the L_{R-I} lagging anchor algorithm becomes

$$\dot{p}_1 = \sum_{j=1}^{2} p_1 p_j (d_1^1 - d_j^1) + (\bar{p}_1 - p_1)$$

$$\dot{\bar{p}}_1 = p_1 - \bar{p}_1$$

$$\dot{q}_1 = \sum_{j=1}^{2} q_1 q_j (d_1^2 - d_j^2) + (\bar{q}_1 - q_1)$$

$$\dot{\bar{q}}_1 = q_1 - \bar{q}_1 \tag{3.61}$$

where $d_1^1 = r_{11} q_1 + r_{12}(1 - q_1)$, $d_2^1 = r_{21} q_1 + r_{22}(1 - q_1)$, $d_1^2 = c_{11} p_1 + c_{21}(1 - p_1)$, and $d_2^2 = c_{12} p_1 + c_{22}(1 - p_1)$. Then (3.61) becomes

$$\dot{p}_1 = p_1(1 - p_1)[u_1 q_1 + r_{12} - r_{22}] + (\bar{p}_1 - p_1)$$

$$\dot{\bar{p}}_1 = p_1 - \bar{p}_1$$

$$\dot{q}_1 = q_1(1 - q_1)[u_2 p_1 + c_{21} - c_{22}] + (\bar{q}_1 - q_1)$$

$$\dot{\bar{q}}_1 = q_1 - \bar{q}_1 \tag{3.62}$$

where $u_1 = r_{11} - r_{12} - r_{21} + r_{22}$ and $u_2 = c_{11} - c_{12} - c_{21} + c_{22}$. If we let the right-hand side of the above equation equal to zero, we then get the equilibrium points of the above equations as $(p_1^*, q_1^*) = (0, 0), (0, 1), (1, 0), (1, 1), ((c_{22} - c_{21})/u_2, (r_{22} - r_{12})/u_1)$. To study the stability of the above learning dynamics, we use a linear approximation of the above equations around the equilibrium point $(p_1^*, q_1^*, p_1^*, q_1^*)$. Then the linearization matrix J is given as

$$J_{(p_1^*, q_1^*)} = \begin{bmatrix} (1 - 2p_1^*)(u_1 q_1^* + \\ r_{12} - r_{22}) - 1 & 1 & p_1^*(1 - p_1^*)u_1 & 0 \\ 1 & -1 & 0 & 0 \\ q_1^*(1 - q_1^*)u_2 & 0 & \begin{matrix}(1 - 2q_1^*)(u_2 p_1^* + \\ c_{21} - c_{22}) - 1\end{matrix} & 1 \\ 0 & 0 & 1 & -1 \end{bmatrix} \tag{3.63}$$

If we substitute each of the equilibrium points $(0,0), (0,1), (1,0), (1,1)$ into (3.63), we get

$$J_{\text{pure}} = \begin{bmatrix} -e_1 - 1 & 1 & 0 & 0 \\ 1 & -1 & 0 & 0 \\ 0 & 0 & -e_2 - 1 & 1 \\ 0 & 0 & 1 & -1 \end{bmatrix} \tag{3.64}$$

where

$$e_1 = r_{22} - r_{12}, e_2 = c_{22} - c_{21} \qquad \text{for } (0,0); \tag{3.65}$$
$$e_1 = r_{21} - r_{11}, e_2 = c_{21} - c_{22} \qquad \text{for } (0,1); \tag{3.66}$$
$$e_1 = r_{12} - r_{22}, e_2 = c_{12} - c_{11} \qquad \text{for } (1,0); \tag{3.67}$$
$$e_1 = r_{11} - r_{21}, e_2 = c_{11} - c_{12} \qquad \text{for } (1,1) \tag{3.68}$$

The eigenvalues of the above matrix J_{pure} are $\lambda_{1,2} = 0.5[-(e_1 + 2) \pm \sqrt{e_1^2 + 4)}]$ and $\lambda_{3,4} = 0.5[-(e_2 + 2) \pm \sqrt{e_2^2 + 4)}]$. In order to obtain a stable equilibrium point, the real parts of the eigenvalues of J_{pure} must be negative. Therefore, the equilibrium point is asymptotically stable if

$$0.5[-(e_{1,2} + 2) \pm \sqrt{e_{1,2}^2 + 4)}] < 0 \quad \Rightarrow$$
$$e_{1,2} + 2 > \sqrt{e_{1,2}^2 + 4} \quad \Rightarrow$$
$$e_{1,2} > 0 \tag{3.69}$$

For the equilibrium point $((c_{22} - c_{21})/u_2, (r_{22} - r_{12})/u_1)$, the linearization matrix becomes

$$J_{\text{mixed}} = \begin{bmatrix} -1 & 1 & p_1^*(1 - p_1^*)u_1 & 0 \\ 1 & -1 & 0 & 0 \\ q_1^*(1 - q_1^*)u_2 & 0 & -1 & 1 \\ 0 & 0 & 1 & -1 \end{bmatrix} \qquad (3.70)$$

The characteristic equation of the above matrix is

$$\lambda^4 + 4\lambda^3 + (4 + e_3)\lambda^2 + 2e_3\lambda + e_3 = 0 \qquad (3.71)$$

where $e_3 = -p_1^*(1 - p_1^*)q_1^*(1 - q_1^*)u_1 u_2$. We set up the Routh table to analyze the locations of the roots in (3.71) as follows:

λ^4	1	$4 + e_3$	e_3
λ^3	4	$2c_3$	
λ^2	$4 + 0.5\,e_3$	e_3	
λ^1	$(e_3^2 + 4e_3)/(4 + 0.5e_3)$		
λ^0	e_3		

(3.72)

Based on the Routh–Hurwitz stability criterion, if (3.71) is stable, then all the coefficients of the equation must be positive and all the elements in the first column of the Routh table in (3.72) are positive. In order to meet the Routh–Hurwitz stability criterion, we must have $e_3 > 0$. Therefore, the equilibrium point $((c_{22} - c_{21})/u_2, (r_{22} - r_{12})/u_1)$ is asymptotically stable if

$$e_3 = -p_1^*(1 - p_1^*)q_1^*(1 - q_1^*)u_1 u_2 > 0 \quad \Rightarrow$$
$$u_1 u_2 < 0 \qquad (3.73)$$
■

Case 3.1 Strict Nash equilibrium in pure strategies We first consider that the game only has strict Nash equilibrium in pure strategies. Without loss of generality, we assume that the Nash equilibrium in this case is both players' first actions. According to the definition of a strict Nash equilibrium in the inequality (3.7), if the Nash equilibrium strategies are both players' first actions, we can get

$$r_{11} > r_{21}, c_{11} > c_{12} \qquad (3.74)$$

Since the Nash equilibrium in this case is the equilibrium point $(1, 1)$, we can get $e_1 = r_{11} - r_{21} > 0$ and $e_2 = c_{11} - c_{12} > 0$ based on (3.68) and (3.74). Therefore, the stability condition (3.69) is satisfied and the equilibrium point $(1, 1)$, which is the Nash equilibrium in this case, is asymptotically stable.

We now test the other equilibrium points. We first consider the equilibrium point $((c_{22} - c_{21})/u_2, (r_{22} - r_{12})/u_1)$. According to the stability condition (3.73), if this equilibrium point is stable, we must have $u_1 u_2 < 0$. To be a valid inner point in the probability space (unit square), the equilibrium point $((c_{22} - c_{21})/u_2, (r_{22} - r_{12})/u_1)$ must satisfy

$$\begin{cases} 0 < (c_{22} - c_{21})/u_2 < 1 \\ 0 < (r_{22} - r_{12})/u_1 < 1 \end{cases} \tag{3.75}$$

If $u_1 u_2 < 0$, we can get

$$\begin{cases} r_{11} > r_{21}, r_{22} > r_{12} \\ c_{11} < c_{12}, c_{22} < c_{21} \end{cases} \quad \text{if } u_1 > 0, u_2 < 0 \tag{3.76}$$

$$\begin{cases} r_{11} < r_{21}, r_{22} < r_{12} \\ c_{11} > c_{12}, c_{22} > c_{21} \end{cases} \quad \text{if } u_1 < 0, u_2 > 0 \tag{3.77}$$

However, the conditions in the inequalities (3.76) and (3.77) conflict with the inequalities in (3.74). Therefore, the inequality $u_1 u_2 < 0$ will not hold and the equilibrium point $((c_{22} - c_{21})/u_2, (r_{22} - r_{12})/u_1)$ is unstable in Case 3.1.

For the equilibrium points $(0, 1)$ and $(1, 0)$, based on (3.66), (3.67), and (3.69), the stability conditions are $r_{21} > r_{11}, c_{21} > c_{22}$ for $(0, 1)$ and $r_{12} > r_{22}, c_{12} > c_{11}$ for $(1, 0)$. However, these stability conditions conflict with the inequalities $r_{11} > r_{21}, c_{11} > c_{12}$ in (3.74). Therefore, the equilibrium points $(0, 1)$ and $(1, 0)$ are unstable in Case 3.1.

For the equilibrium point $(0, 0)$, the stability condition is $r_{22} > r_{12}, c_{22} > c_{21}$ based on conditions (3.65) and (3.69). From the inequality (3.7), we can find that this stability condition also meets the requirement for a strict NE (both players' second actions) in inequality (3.7). Therefore, the equilibrium point $(0, 0)$ is stable only if it is also an NE point.

Thus, the NE point is asymptotically stable, while any equilibrium point which is not an NE is unstable.

Case 3.2 Nash equilibrium in fully mixed strategies We now consider that the game only has Nash equilibrium in fully mixed strategies. Singh et al. [9] showed that a Nash equilibrium in fully mixed strategies for a two-player two-action general-sum matrix game has the form

$$(p_1^{NE}, q_1^{NE}) = \left[\frac{c_{22} - c_{21}}{u_2}, \frac{r_{22} - r_{12}}{u_1} \right] \tag{3.78}$$

where (p_1^{NE}, q_1^{NE}) denotes the Nash equilibrium strategies over players' first actions which happens to be the equilibrium point of (3.62). According to the condition (3.73), the equilibrium point $((c_{22} - c_{21})/u_2, (r_{22} - r_{12})/u_1)$ is asymptotically stable if $u_1 u_2 < 0$. If we assume $u_1 u_2 > 0$, we can get

$$\begin{cases} 0 < (c_{22} - c_{21})/u_2 < 1 \\ 0 < (r_{22} - r_{12})/u_1 < 1 \end{cases}$$

$$\begin{cases} r_{11} > r_{21}, r_{22} > r_{12} \\ c_{11} > c_{12}, c_{22} > c_{21} \end{cases} \quad \text{if } u_1 > 0, u_2 > 0 \tag{3.79}$$

$$\begin{cases} r_{11} < r_{21}, r_{22} < r_{12} \\ c_{11} < c_{12}, c_{22} < c_{21} \end{cases} \quad \text{if } u_1 < 0, u_2 < 0 \tag{3.80}$$

According to the inequality (3.7), the above equations contain multiple NEs in pure strategies: $(p_1^{NE}, q_1^{NE}) = (1, 1), (0, 0)$ if $u_1 > 0, u_2 > 0$ and $(p_1^{NE}, q_1^{NE}) = (0, 1), (1, 0)$ if $u_1 < 0, u_2 < 0$. However, under our assumption, the game in Case 3.2 only has an NE in fully mixed strategies, and NEs in pure strategies do not exist. Therefore, we always have $u_1 u_2 < 0$ in Case 3.2 and the equilibrium point $((c_{22} - c_{21})/u_2, (r_{22} - r_{12})/u_1)$, which is also the NE point, is asymptotically stable.

For the other equilibrium points, based on conditions (3.65)–(3.68) and (3.69), the stability conditions become

$$r_{22} > r_{12}, c_{22} > c_{21} \quad \text{for } (0, 0) \tag{3.81}$$

$$r_{21} > r_{11}, c_{21} > c_{22} \quad \text{for } (0, 1) \tag{3.82}$$

$$r_{12} > r_{22}, c_{12} > c_{11} \quad \text{for } (1, 0) \tag{3.83}$$

$$r_{11} > r_{21}, c_{11} > c_{12} \quad \text{for } (1, 1) \tag{3.84}$$

As already noted, the game in Case 3.2 only has an NE in fully mixed strategies and we always have $u_1 u_2 < 0$. Then the inequalities (3.76) and (3.77) are true in Case 3.2. However, the stability conditions (3.81)–(3.84) for the equilibrium points $(0, 0), (0, 1), (1, 0), (1, 1)$ conflict with the inequalities (3.76) and (3.77). Therefore, the equilibrium points other than $((c_{22} - c_{21})/u_2, (r_{22} - r_{12})/u_1)$ are unstable in this case.

Thus we can conclude that the NE point is asymptotically stable while the other equilibrium points are unstable in Case 3.2.

3.14.1 Simulation

We now simulate three matrix games to show the performance of the L_{R-I} lagging anchor algorithm. The first game is the matching pennies game. This game is a two-player zero-sum game and each player has two actions: heads or tails. If both players choose the same action, then player 1 gets a reward 1 and player 2 gets a reward -1. If the actions are different, then player 1 gets -1 and player 2 gets 1. Based on the reward matrix in Table 3.1(a) and the solutions in Example 3.1, the NE in this game is in fully mixed strategies such that each player plays heads and tails with a probability of 0.5. We set the step size $\eta = 0.001$ in (3.58) and $p_1(0) = q_1(0) = 0.2$. We run the simulation for 30,000 iterations. In Fig. 3-9, the players' probabilities of taking their first actions start from (0.2, 0.2) and move close to the NE point (0.5, 0.5) as the learning proceeds.

The second game we simulate is a two-player general-sum game, namely the prisoners' dilemma. In this game, we have two players and each player has two actions: defect or cooperate. A player receives a reward of 10 if it defects and the other player cooperates, or receives a reward of 0 if it cooperates and the other player defects. If both players cooperate, each player receives a reward of 5. If they both defect, each player receives a reward of 1. The reward matrix is shown in Table 3.3(b), where one player's reward matrix is the transpose of the other player's reward matrix. This game has a unique NE in pure strategies which is both players playing defect. We set the step size $\eta = 0.001$ in (3.58) and $p_1(0) = q_1(0) = 0.5$. We run the simulation for 30,000 iterations. Figure 3-10 shows that the players' strategies move close to the NE strategies (both players' second actions) as the learning proceeds.

Table 3.3 Examples of two-player matrix games.

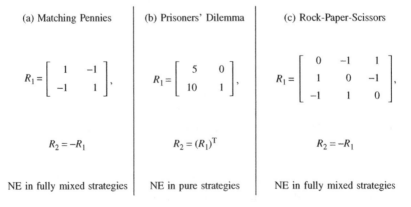

(a) Matching Pennies	(b) Prisoners' Dilemma	(c) Rock-Paper-Scissors
$R_1 = \begin{bmatrix} 1 & -1 \\ -1 & 1 \end{bmatrix}$,	$R_1 = \begin{bmatrix} 5 & 0 \\ 10 & 1 \end{bmatrix}$,	$R_1 = \begin{bmatrix} 0 & -1 & 1 \\ 1 & 0 & -1 \\ -1 & 1 & 0 \end{bmatrix}$,
$R_2 = -R_1$	$R_2 = (R_1)^{\mathrm{T}}$	$R_2 = -R_1$
NE in fully mixed strategies	NE in pure strategies	NE in fully mixed strategies

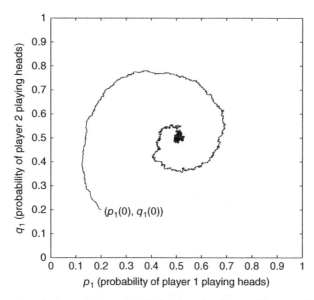

Fig. 3-9. Trajectories of players' strategies during learning in matching pennies. Reproduced from [8], © X. Lu.

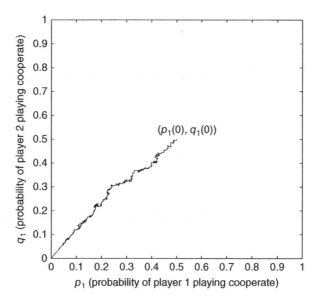

Fig. 3-10. Trajectories of players' strategies during learning in prisoners' dilemma. Reproduced from [8], © X. Lu.

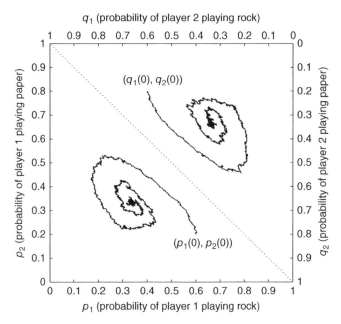

Fig. 3-11. Trajectories of players' strategies during learning in rock-paper-scissors.
Reproduced from [8], © X. Lu.

The third game we simulate in this chapter is the rock-paper-scissors game. This game has two players and each player has three actions: rock, paper, and scissors. A winner in the game is determined by the following rules: paper defeats rock, scissors defeats paper, and rock defeats scissors. The winner receives a reward of 1 and the loser receives -1. If both players choose the same action, each player gets 0. The reward matrix is shown in Table 3.3(c). This game has an NE in fully mixed strategies which is each player choosing any action with the same probability of $1/3$. We set the step size $\eta = 0.001$, $p_1(0) = q_1(0) = 0.6$, and $p_2(0) = q_2(0) = 0.2$. We run the simulation for 50,000 iterations. Although we only prove the convergence for two-player two-action games, the result in Fig. 3-11 shows that the proposed L_{R-I} lagging anchor algorithm may be applicable to a two-player matrix game with more than two actions.

References

[1] M. J. Osborne and A. Rubinstein, *A Course in Game Theory*. Cambridge, Massachusetts: The MIT Press, 1994.

[2] G. Owen, *Game Theory*. San Diego, California: Academic Press, 1995.

[3] M. Bowling and M. Veloso, "Multiagent learning using a variable learning rate," *Artificial Intelligence*, vol. 136, no. 2, pp. 215–250, 2002.

[4] T. Başar and G. J. Olsder, *Dynamic Noncooperative Game Theory, SIAM Series in Classics in Applied Mathematics*. London: Academic Press, 2nd ed., 1999.

[5] P. Sastry, V. Phansalkar, and M. Thathachar, "Decentralized learning of Nash equilibria in multi-person stochastic games with incomplete information," *IEEE Transactions on Systems, Man, and Cybernetics*, vol. 24, no. 5, pp. 769–777, 1994.

[6] M. J. Osborne, *An Introduction to Game Theory*. Oxford: Oxford University Press, 2003.

[7] M. L. Littman, "Markov games as a framework for multi-agent reinforcement learning," in 11th International Conference on Machine Learning, (New Brunswick, United States), July 1994, pp. 157–163, 1994.

[8] X. Lu, Ph.D., "On Multi-Agent Reinforcement Learning in Games." Ph.D. Thesis, Carleton University, Ottawa, ON, 2012.

[9] S. P. Singh, M. J. Kearns, and Y. Mansour, "Nash convergence of gradient dynamics in general-sum games," in UAI '00: Proceedings of the 16th Conference in Uncertainty in Artificial Intelligence, Stanford University, Stanford, California, USA, June 30 - July 3, 2000, pp. 541–548, 2000.

[10] C. J. C. H. Watkins and P. Dayan, "Q-learning," *Machine Learning*, vol. 8, no. 3, pp. 279–292, 1992.

[11] E. Yang and D. Gu, "A survey on multiagent reinforcement learning towards multi-robot systems," in Proceedings of IEEE Symposium on Computational Intelligence and Games, 2005.

[12] L. Buşoniu, R. Babuška, and B. D. Schutter, "Multiagent reinforcement learning: a survey," 9th International Conference on Control, Automation, Robotics and Vision (ICARCV), pp. 1–6, 2006.

[13] X. Lu and H. M. Schwartz, "An investigation of guarding a territory problem in a grid world," in American Control Conference, pp. 3204–3210, 2010.

[14] S. Lakshmivarahan and K. S. Narendra, "Learning algorithms for two-person zero-sum stochastic games with incomplete information," *Mathematics of Operations Research*, vol. 6, no. 3, pp. 379–386, 1981.

[15] S. Lakshmivarahan and K. S. Narendra, "Learning algorithms for two-person zero-sum stochastic games with incomplete information: a unified approach," *SIAM Journal on Control and Optimization*, vol. 20, no. 4, pp. 541–552, 1982.

[16] F. A. Dahl, "The lagging anchor algorithm: reinforcement learning in two-player zero-sum games with imperfect information," *Machine Learning*, vol. 49, pp. 5–37, 2002.

[17] F. A. Dahl, "The lagging anchor model for game learning—a solution to the crawford puzzle," *Journal of Economic Behavior & Organization*, vol. 57, pp. 287–303, 2005.

[18] K. S. Narendra and M. A. L. Thathachar, *Learning Automata: An Introduction*. Englewood Cliffs, New Jersey: Prentice Hall, 1989.

[19] M. Thathachar and P. Sastry, *Networks of Learning Automata: Techniques for Online Stochastic Optimization*. Boston, Massachusetts: Kluwer Academic Publishers, 2004.

Chapter 4
Learning in Multiplayer Stochastic Games

4.1 Introduction

The agents in a multiagent system can be to some degree preprogrammed with behaviors designed in advance. It is often necessary that the agents be able to learn online such that the performance of the multiagent system improves. However, typically a multiagent system is very complex and preprogramming the system is for practical reasons impossible. Furthermore, the dynamics of the agents and the environment can change over time and learning and adaptation is required.

In early work on multiagent reinforcement learning (MARL) for stochastic games [1], it was recognized that no agent works in a vacuum. In his seminal paper, Littman [1] focused on only two agents that had opposite and opposing goals. This means that they could use a single reward function which one tried to maximize and the other tried to minimize. The agent had to work with a competing agent and had to behave so as to maximize their reward in the worst possible case. They also recognized the need for mixed strategies because the agent or player could not be certain of the action taken by its opponent. Littman [1] introduced the minimax Q-learning algorithm. We have already shown the idea of the minimax Q-learning algorithm in Chapter 3, Section 3.2.

Multi-Agent Machine Learning: A Reinforcement Approach, First Edition. Howard M. Schwartz.
© 2014 John Wiley & Sons, Inc. Published 2014 by John Wiley & Sons, Inc.

In a rational multiagent game, each agent must keep track in some way of what the other learning agents are doing. The types of games and situations that the learning agent may encounter include fully competitive games that are zero-sum games. There are also general-sum cooperative games, where the agents try to get the maximum reward by cooperating with each other. For example, in the prisoners' dilemma problem, if it is a cooperative game, then each player should lie to the police and cooperate with each other and get the least time in jail. However, if it is a competitive game, then they should each defect to mitigate against the worst case scenario where one lies to police and the opponent confesses and the first one goes to jail for life. However, the agents must communicate to cooperate with each other.

Typically, in a multiagent system, the agents must keep track of the other agents' behaviors such that a coherent behavior results. Furthermore, we have the issue of scalability to consider. The agents must keep track of a large number of possible states and joint actions.

Learning in stochastic games can be formalized as a MARL problem [2]. Agents select actions simultaneously at the current state and receive rewards at the next state. Different from the algorithm that can solve for a Nash equilibrium in a stochastic game, the goal of a reinforcement learning algorithm is to learn equilibrium strategies through interaction with the environment. Generally, in a MARL problem, agents may not know the transition function or the reward function from the environment. Instead, agents are required to select actions and observe the received reward and the next state in order to gain information of the transition function or the reward function.

Rationality and convergence are two desirable properties for multiagent learning algorithms in stochastic games [2]. When we say that a player is rational, we mean that, if the other players' policies converge to stationary policies, then the learning algorithm for will converge to a policy that is a best response to the other players' policies. What does this mean? Let us say that you are playing the matching pennies game against a *bad* player who always plays heads. You win whenever you both play heads or tails. At first, you assume that your opponent is a rational player and, therefore, you assume that your opponent is playing the rational strategy of playing heads and tails each 50% of the time. Therefore, you start out playing heads and tails 50% of the time. However, after a number of plays you realize that your opponent always seems to be playing heads as his stationary strategy. You would then quickly shift and begin to also play heads all the time. Then you always win and you are playing the rational strategy but your opponent is not playing a rational strategy.

In the stochastic learning algorithms, we also have the idea of convergence. Let us say that all the other players are playing a stationary strategy. They are not learning or changing their strategy in any way. Then you adapt to this behavior and converge to some rational strategy. Or, let us say that everyone in the game is adapting according to the same algorithm. Then do all the players converge to an optimal strategy or a Nash equilibrium? If all the players use rational learning algorithms and their policies converge, then they must have converged to an equilibrium. Each player will play the best response to all other players.

In this chapter, we review some existing reinforcement learning algorithms in stochastic games. We analyze these algorithms based on their applicability, rationality, and convergence properties.

Isaacs [3] introduced a differential game of guarding a territory where a defender tries to intercept an invader before the invader reaches the territory. In this chapter, we introduce a grid version of Isaacs' game called the *grid game of guarding a territory*. It is a two-player zero-sum stochastic game where the defender plays against the invader in a grid world. We then study how the players learn to play the game using MARL algorithms. We apply two reinforcement learning algorithms to this game and test the performance of these learning algorithms based on the convergence and rationality properties.

4.2 Multiplayer Stochastic Games

A Markov decision process contains a single player and multiple states, whereas a matrix game contains multiple players and a single state. For a game with more than one player and multiple states, we define a stochastic game (or Markov game) as the combination of Markov decision processes and matrix games. A stochastic game is a tuple $(n, S, A_1, \ldots, A_n, T, \gamma, R_1, \ldots, R_n)$ where n is the number of the players, $T : S \times A_1 \times \cdots \times A_n \times S \to [0, 1]$ is the transition function, $A_i (i = 1, \ldots, n)$ is the action set for the player i, $\gamma \in [0, 1]$ is the discount factor, and $R_i : S \times A_1 \times \cdots \times A_n \times S \to \mathbb{R}$ is the reward function for player i. The transition function in a stochastic game is a probability distribution over next states given the current state and joint action of the players. The reward function $R_i(s, a_1, \ldots, a_n, s')$ denotes the reward received by player i in state s' after taking joint action $(a_1, \ldots, a_n)$ in state s. Similar to Markov decision processes, stochastic games also have the Markov property. That is, the player's next state and reward depend only on the current state and all the players' current actions.

For a multiplayer stochastic game, we want to find the Nash equilibria in the game if we know the reward function and transition function in the game. A Nash equilibrium in a stochastic game can be described as a tuple of n strategies $(\pi_1^*, \dots, \pi_n^*)$ such that for all $s \in S$ and $i = 1, \dots, n$,

$$V_i(s, \pi_1^*, \dots, \pi_i^*, \dots, \pi_n^*) \geq V_i(s, \pi_1^*, \dots, \pi_i, \dots, \pi_n^*) \text{ for all } \pi_i \in \Pi_i \quad (4.1)$$

where Π_i is the set of strategies available to player i, and $V_i(s, \pi_1^*, \dots, \pi_n^*)$ is the expected sum of discounted rewards for player i given the current state and all the players' equilibrium strategies. To simplify notation, we use $V_i^*(s)$ to represent $V_i(s, \pi_1^*, \dots, \pi_n^*)$ as the state-value function under Nash equilibrium strategies. We can also define the action-value function $Q^*(s, a_1, \dots, a_n)$ as the expected sum of discounted rewards for player i given the current state and the current joint action of all the players, and following the Nash equilibrium strategies thereafter. Then we can get

$$V_i^*(s) = \sum_{a_1, \dots, a_n \in A_1 \times \cdots \times A_n} Q_i^*(s, a_1, \dots, a_n)\pi_1^*(s, a_1) \cdots \pi_n^*(s, a_n) \quad (4.2)$$

$$Q_i^*(s, a_1, \dots, a_n) = \sum_{s' \in S} T(s, a_1, \dots, a_n, s')[R_i(s, a_1, \dots, a_n, s') + \gamma V_i^*(s')] \quad (4.3)$$

where $\pi_i^*(s, a_i) \in PD(A_i)$ is a probability distribution over action a_i under player i's Nash equilibrium strategy, $T(s, a_1, \dots, a_n, s') = \Pr\{s_{k+1} = s' | s_k = s, a_1, \dots, a_n\}$ is the probability of the next state being s' given the current state s and joint action $(a_1, \dots, a_n)$, and $R_i(s, a_1, \dots, a_n, s')$ is the expected immediate reward received in state s' given the current state s and joint action $(a_1, \dots, a_n)$. Based on (4.2) and (4.3), the Nash equilibrium in (4.1) can be rewritten as

$$\sum_{a_1, \dots, a_n \in A_1 \times \cdots \times A_n} Q_i^*(s, a_1, \dots, a_n)\pi_1^*(s, a_1) \cdots \pi_i^*(s, a_i) \cdots \pi_n^*(s, a_n)$$

$$\geq \sum_{a_1, \dots, a_n \in A_1 \times \cdots \times A_n} Q_i^*(s, a_1, \dots, a_n)\pi_1^*(s, a_1) \cdots \pi_i(s, a_i) \cdots \pi_n^*(s, a_n) \quad (4.4)$$

Stochastic games can be classified on the basis of the players' reward functions. If all the players have the same reward function, the game is called a *fully cooperative game* or a *team game*. If one player's reward function has always the opposite sign of the other player's, the game is called a *two-player fully competitive game* or *zero-sum game*. For the game with all types of reward functions, we call it a general-sum stochastic game.

To solve a stochastic game, we need to find a strategy $\pi_i : S \rightarrow A_i$ that can maximize player i's discounted future reward with a discount factor γ. Similar to the strategy in matrix games, the player's strategy in a stochastic game is probabilistic. An example is the soccer game introduced by Littman [1], where an agent on the offensive side must use a probabilistic strategy to pass an unknown defender. In the literature, a solution to a stochastic game is described as Nash equilibrium strategies in a set of associated *state-specific matrix games* [1, 4]. A state-specific matrix game is also called a *stage game*. In these state-specific matrix games, we define the action-value function $Q_i^*(s, a_1, \ldots, a_n)$ as the expected reward for player i when all the players take joint action $a_1, \ldots, a_n$ in state s and follow the Nash equilibrium strategies thereafter. If the value of $Q_i^*(s, a_1, \ldots, a_n)$ is known for all the states, we can find player i's Nash equilibrium strategy by solving the associated state-specific matrix game [4]. Therefore, for each state s, we have a matrix game and we can find the Nash equilibrium strategies in this matrix game. Then the Nash equilibrium strategies for the game are the collection of Nash equilibrium strategies in each state-specific matrix game for all the states. We present an example here.

Example 4.1 We define a 2×2 grid game with two players denoted as $P1$ and $P2$. Two players' initial positions are located at the bottom left corner for player 1 and the upper right corner for player 2, as shown in Fig. 4-1a. Both players try to reach one of the two goals denoted as "G" in minimum number of steps. Starting from their initial positions, each player has two possible moves which are moving up or right for player 1, and moving left or down for player 2. Figure 4-1b shows the numbered cells in this game. Each player takes an action and moves one cell at a time. The game ends when either of the players reaches the goal and receives a reward 10. The dashed line between the upper cells and the bottom cells in Figure 4-1a is the barrier that the player can pass through with a probability 0.5. If both players move to the same cell, both players bounce back to their original positions. Figure 4-1c shows the possible transitions in the game. The number of possible states (players' joint positions) is seven containing the players' initial positions $s_1 = (2, 3)$ and six terminal states $(s_2, \ldots, s_7)$.

According to the above description of the game, we can find the Nash equilibrium strategies in this example. The players need to avoid the barrier and move to the goals next to them without crossing the barrier. Therefore, the Nash equilibrium is the players' joint action $(a_1 = \text{Right}, a_2 = \text{Left})$. Based on (4.2) and (4.3), the state-value function $V_i^*(s_1)$ under the Nash equilibrium strategies is

$$V_i^*(s_1) = R_i(s_1, \text{Right, Left}, s_7) + \gamma V_i^*(s_7)$$

$$= 10 + 0.9 \cdot 0 = 10 \tag{4.5}$$

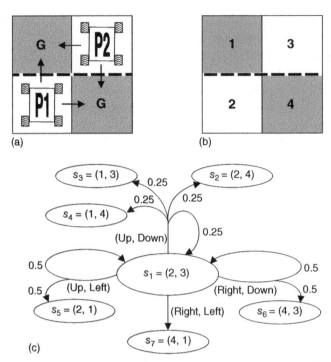

Fig. 4-1. Example of stochastic games. (a) A 2 × 2 grid game with two players. (b) The numbered cells in the game. (c) Possible state transitions given players' joint action (a_1, a_2). Reproduced from [5], © X. Lu.

where $\gamma = 0.9$, $R_i(s_1, \text{Right}, \text{Left}, s_7) = 10$, and $V_i^*(s_7) = 0$ (the state-value functions at terminal states are always zero). We can also find the action-value function $Q_i^*(s_1, a_1, a_2)$. For example, the action-value function $Q_1^*(s_1, \text{Up}, \text{Down})$ for player 1 can be written as

$$Q_1^*(s_1, \text{Up}, \text{Down}) = \sum_{s' = s_1 \sim s_4} T(s_1, \text{Up}, \text{Down}, s')[R_1(s_1, \text{Up}, \text{Down}, s') + \gamma V_1^*(s')]$$

$$= 0.25(0 + 0.9 V_1^*(s_1)) + 0.25(0 + 0.9 V_1^*(s_2))$$

$$+ 0.25(10 + 0.9 V_1^*(s_3)) + 0.25(10 + 0.9 V_1^*(s_4))$$

$$= 0.25 \cdot 0.9 \cdot 10 + 0.25 \cdot 0 + 0.25 \cdot 10 + 0.25 \cdot 10$$

$$= 7.25 \tag{4.6}$$

Table 4.1 shows the action-value functions under the players' Nash equilibrium strategies.

Table 4.1 Action-value function $Q_i^*(s_1, a_1, a_2)$ in Example 4.1.

	a_2				a_2	
$Q_1^*(s_1, a_1, a_2)$	Left	Down		$Q_2^*(s_1, a_1, a_2)$	Left	Down
Up	4.5	7.25		Up	9.5	7.25
Right	10	9.5		Right	10	4.5

In the left table a_1 labels the rows; in the right table a_1 labels the rows.

4.3 Minimax-Q Algorithm

Littman [1] proposed a minimax-Q algorithm specifically designed for two-player zero-sum stochastic games. The minimax-Q algorithm uses the minimax principle to solve for players' Nash equilibrium strategies and values of states for two-player zero-sum stochastic games. Similar to Q-learning, minimax-Q algorithm is a temporal-difference learning method that performs backpropagation on values of states or state-action pairs. We show the minimax-Q algorithm as follows:

In a two-player zero-sum stochastic game, given the current state s, we define the state-value function for player i as

$$V_i^*(s) = \max_{\pi_i(s,\cdot)} \min_{a_{-i} \in A_{-i}} \sum_{a_i \in A_i} Q_i^*(s, a_i, a_{-i})\pi_i(s, a_i), \quad (i = 1, 2) \qquad (4.7)$$

where $-i$ denotes player i's opponent, $\pi_i(s, \cdot)$ denotes all the possible strategies of player i at state s, and $Q_i^*(s, a_i, a_{-i})$ is the expected reward when player i and its opponent choose action $a_i \in A_i$ and $a_{-i} \in A_{-i}$, respectively, and follow their Nash equilibrium strategies after that. If we know $Q_i^*(s, a_i, a_{-i})$, we can solve Eq. (4.7) and find player i's Nash equilibrium strategy $\pi^*(s)$. Similar to finding the minimax solution for (3.8), one can use linear programming to solve Eq. (4.7). For a MARL problem, $Q_i^*(s, a_i, a_{-i})$ is unknown to the players in the game.

The minimax-Q algorithm is listed in Algorithm 4.1. The minimax-Q algorithm can guarantee the convergence to a Nash equilibrium if all the possible states and players' possible actions are visited infinitely often. The proof of convergence for the minimax-Q algorithm can be found in Reference 6. One drawback of this algorithm is that we have to use linear programming to solve for $\pi_i(s)$ and $V_i(s)$ at each iteration in Algorithm 4.1.

Algorithm 4.1 Minimax-Q algorithm

1: Initialize $Q_i(s, a_i, a_{-i})$, $V_i(s)$ and π_i
2: **for** Each iteration **do**
3:   Player i takes an action a_i from current state s based on an exploration-exploitation strategy
4:   At the subsequent state s', player i observes the received reward r_i and the opponent's action taken at the previous state s.
5:   Update $Q_i(s, a_i, a_{-i})$:

$$Q_i(s, a_i, a_{-i}) \leftarrow (1 - \alpha)Q_i(s, a_i, a_{-i}) + \alpha\left[r_i + \gamma V_i(s')\right] \qquad (4.8)$$

where α is the learning rate and γ is the discount factor.
6:   Use linear programming to solve Eq. (4.7) and obtain the updated $\pi_i(s)$ and $V_i(s)$
7: **end for**

This will lead to a slow learning process. Also, in order to perform linear programming, the player i has to know the opponent's action space.

Using the minimax-Q algorithm, the player will always play a "safe" strategy in case of the worst scenario caused by the opponent. However, if the opponent is currently playing a stationary strategy which is not its equilibrium strategy, the minimax-Q algorithm cannot make the player adapt its strategy to the change in the opponent's strategy. The reason is that the minimax-Q algorithm is an opponent-independent algorithm and it will converge to the player's Nash equilibrium strategy no matter what strategy the opponent uses. If the player's opponent is a weak opponent that does not play its equilibrium strategy, then the player's optimal strategy is not the same as its Nash equilibrium strategy. The player's optimal strategy will do better than the player's Nash equilibrium strategy in this case.

Overall, the minimax-Q algorithm, which is applicable to zero-sum stochastic games, does not satisfy the rationality property but satisfies the convergence property. The following example of a 2×2 grid game demonstrates the algorithm.

4.3.1 2 × 2 Grid Game

The playing field of the 2×2 grid game is shown in Fig. 4-2. The territory to be guarded is located at the bottom-right corner. Initially, the invader starts at the top-left corner while the defender starts at the same cell as the territory. To better illustrate the guarding a territory problem, we simplify the possible number of actions of each player to 2. The invader can only move down or right, while the defender can only move up or left. The capture of the invader happens when the defender and the invader move into the same cell excluding the territory cell. The game ends when the invader reaches the territory or the defender catches the invader before it reaches the territory. We suppose both players start from the initial state s_1, as shown in Fig. 4-2a. There are three nonterminal states (s_1, s_2, s_3) in this game shown in Fig. 4-2. If the invader moves to the right cell and the defender happens to move left, then both players reach the state s_2 in Fig. 4-2b. If the invader moves down and the defender moves up simultaneously, then they will reach the state s_3 in Fig. 4-2c. In states s_2 and s_3, if the invader is smart enough, it can always reach the territory no matter what action the defender will take. As such, the game only has one step because, if the invader gets to state s_2 or s_3, it de facto wins. Therefore, starting from the initial state s_1, a clever defender will try to intercept the invader by guessing which direction the invader will go.

We define the reward functions for the players. The reward function for the defender is defined as

$$
R_D = \begin{cases} dist_{IT}, & \text{defender captures the invader} \\ -10, & \text{invader reaches the territory} \end{cases} \tag{4.9}
$$

where

$$
dist_{IT} = |x_I(t_f) - x_T| + |y_I(t_f) - y_T|
$$

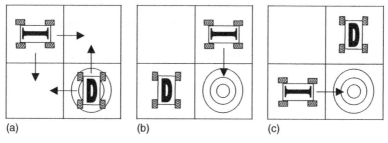

(a)　　　　　　　(b)　　　　　　　(c)

Fig. 4-2. A 2 × 2 grid game. (a) Initial positions of the players: state s_1. (b) Invader in top-right versus defender in bottom-left: state s_2. (c) Invader in bottom-left versus defender in top-right: state s_3. Reproduced from [5], © X. Lu.

The reward function for the invader is given by

$$R_I = \begin{cases} -dist_{IT}, & \text{defender captures the invader} \\ 10, & \text{invader reaches the territory} \end{cases} \tag{4.10}$$

The reward functions (4.9) and (4.10) are also used in the 6×6 grid game.

Before the simulation, we can simply solve this game similar to solving Example 4.1. In the states s_2 and s_3, a smart invader will always reach the territory without being intercepted. The value of the states s_2 and s_3 for the defender will be $V_D(s_2) = -10$ and $V_D(s_3) = -10$. We set the discount factor as 0.9 and we can get $Q_D^*(s_1, a_{left}, o_{right}) = \gamma V_D(s_2) = -9$, $Q_D^*(s_1, a_{up}, o_{down}) = \gamma V_D(s_3) = -9$, $Q_D^*(s_1, a_{left}, o_{down}) = 1$, and $Q_D^*(s_1, a_{up}, o_{right}) = 1$, as shown in Table 4.2 and Table 4.3a. Under the Nash equilibrium, we define the probabilities of the defender moving up and left as $\pi_D^*(s_1, a_{up})$ and $\pi_D^*(s_1, a_{left})$, respectively. The probabilities of the invader moving down and right are denoted as $\pi_I^*(s_1, o_{up})$ and $\pi_I^*(s_1, o_{left})$, respectively. Based on the Q-values in Table 4.3a, we can find the value of the state s_1 for the defender by solving a linear programming problem shown in Table 4.3b. The approach for solving a linear programming problem can be found in Section 3.2.

We define $R(s, a, o)$ to denote the immediate reward to the agent for taking action $a \in A$ in state $s \in S$ when its opponents take action $o \in O$. The idea is for the agent to maximize its reward in the worst possible case. In other words, the agent assumes that the other agent is playing its best possible strategy in a purely competitive game. Therefore, the agent's optimal policy would be to maximize the minimum possible reward. Just like in the case of the prisoners' dilemma, the action should be to defect because this will limit cost of the worst possible outcome regardless of what the other player does. If the prisoner/player decides not to defect, and if the other player does defect, then the prisoner/player will go to jail for a long time; therefore, to maximize the worst possible outcome, the prisoner/player should defect.

Now imagine that we would be satisfied with a policy that guarantees an expected reward $\overline{R}$ no matter what the other player, in this case an opponent, chooses as its action. In this case, the defender's reward matrix in state s_1 is

$$R_D(s_1) = \begin{bmatrix} -9 & 1 \\ 1 & -9 \end{bmatrix} \tag{4.11}$$

Recall that the policy in this case is given as the probability that the agent in state s_1 will choose to go up, $\pi_D(s_1, a_{up})$, and the probability that the agent will choose to go left, $\pi_D(s_1, a_{left})$.

Table 4.2   Minimax solution for the defender in the state s_1.

		Defender	
	Q_D^*	Up	Left
Invader	Down	-9	1
	Right	1	-9

Then we get the following equations for the expected reward, regardless of which action the opponent takes:

$$(-9) \cdot \pi_D(s_1, a_{up}) + (1) \cdot \pi_D(s_1, a_{left}) = \bar{R}$$
$$(1) \cdot \pi_D(s_1, a_{up}) + (-9) \cdot \pi_D(s_1, a_{left}) = \bar{R}$$
$$\pi_D(s_1, a_{up}) + \pi_D(s_1, a_{left}) = 1$$

Therefore, the goal is to maximize the expected reward $\bar{R}$ regardless of what the opponent does.

After solving the linear constraints in Table 4.3b, we get the value of the state s_1 for the defender as $V_D(s_1) = -4$ and the Nash equilibrium strategy for the defender as $\pi_D^*(s_1, a_{up}) = 0.5$ and $\pi_D^*(s_1, a_{left}) = 0.5$. For a two-player zero-sum game, we can get $Q_D^* = -Q_I^*$. Similar to the approach in Table 4.3, we can find the minimax solution of this game for the invader as $V_I(s_1) = 4$, $\pi_I^*(s_1, o_{down}) = 0.5$, and $\pi_I^*(s_1, o_{right}) = 0.5$. Therefore, the Nash equilibrium strategy of the invader is to move down or right with probability 0.5, and the Nash equilibrium strategy of the defender is to move up or left with probability 0.5.

Table 4.3   Minimax solution for the defender in the state s_1. (a) Q-values of the defender for the state s_1. (b) Linear constraints for the defender in the state s_1.

		Defender		Objective: Maximize $\bar{R}$
	Q_D^*	Up	Left	$(-9) \cdot \pi_D(s_1, a_{up}) + (1) \cdot \pi_D(s_1, a_{left}) \geq \bar{R}$
Invader	Down	-9	1	$(1) \cdot \pi_D(s_1, a_{up}) + (-9) \cdot \pi_D(s_1, a_{left}) \geq \bar{R}$
	Right	1	-9	$\pi_D(s_1, a_{up}) + \pi_D(s_1, a_{left}) = 1$
(a)				(b)

The implementation of the minimax Q-learning algorithm begins by initializing the Q matrix, the value function $V(s)$, and the policy $\pi_i(a)$ for each agent. In the example shown here, we initialize $Q(s, a_i, a_{-i}) = 0$ and $V_i(s) = 0$. We arbitrarily initialized the probability of the defender to go up as $\pi_d(up) = 1.0$ and $\pi_d(left) = 0.0$ and of the invader $\pi_i(right) = 1.0$ and $\pi_i(down) = 0.0$. We set the discount factor as $\gamma = 0.9$, the learning rate to $\alpha = 0.1$, and the exploration probability to $\varepsilon = 0.1$. The constraint condition for the linear programming problem for the defender is

$$R_D(s, up, right)\pi(s_1, up) + R_D(s, left, right)\pi(s_1), left) \geq V_D(s_1)$$

$$R_D(s, up, down)\pi(s_1, up) + R_D(s, left, down)\pi(s_1), left) \geq V_D(s_1)$$

$$\pi_D(s_1, up) + \pi_D(s_1, left) = 1 \tag{4.12}$$

Similarly, for the invader we get

$$R_I(s, right, up)\pi_I(s_1, right) + R_I(s, down, up)\pi_I(s_1), down) \geq V_I(s_1)$$

$$R_I(s, right, left)\pi_I(s_1, right) + R_I(s, down, left)\pi_I(s_1), down) \geq V_I(s_1)$$

$$\pi_I(s_1, right) + \pi_I(s_1, down) = 1 \tag{4.13}$$

The linear programming algorithm is run separately for both the invader and the defender. The algorithm (such as the simplex method) will determine the values of $\pi_D(s_1, up)$, $\pi_D(S_1, left)$, and $V_D(s_1)$ at each iteration. The agents do not know a priori the rewards. The best estimate of the rewards is the state-action function $Q(s, a)$. Therefore, the minimax Q-learning algorithm updates both the expected rewards and the policy, $\pi(s, a)$, simultaneously. To use MATLAB to compute the linear programming solution, we need to get the equations into the correct form. MATLAB structures the linear programming problem as

$$\min_x f^T x \tag{4.14}$$

given the constraints

$$A \cdot x \leq b \quad \text{inequality constraints}$$

$$A_{eq} \cdot x = b_{eq} \quad \text{equality constraints}$$

and

$$lb \leq x \leq ub \text{ lower and upper bounds on x}$$

Notice that MATLAB formulates the problem as a minimization problem. We convert the reward maximization problem in the minimax algorithm into a minimization problem by multiplying by -1. We substitute the agents' best estimate of its reward for $R_D(s_1, a_i, a_{-i})$ and $R_I(s_1, a_i, a_{-i})$ with $Q_D(s_1, a_i, a_{-i})$ and $Q_I(s_1, a_i, a_{-i})$ and rewrite the minimization problem for the defender and the invader. For the case of the defender, we write the minization problem as

$$\min_{x_d} f_d^T x_d$$

where $f_D^T = [0, 0, -1]$ and $x_d^T = [\pi_D(s_1, up), \pi_D(s_1, left), V_D(s_1)]$, subject to the constraint

$$-Q_D(s_1, up, right)\pi_D(s_1, up) - Q_D(s_1, left, right)\pi_D(s_1, left) + V_D(s_1) \le 0$$
$$-Q_D(s_1, up, down)\pi_D(s_1, up) - Q_D(s_1, left, down)\pi_D(s_1, left) + V_D(s_1) \le 0$$

and

$$\pi_D(s_1, up) + \pi_D(s_1, left) = 1$$

Then the matrix A becomes

$$A = \begin{bmatrix} -Q_D(s_1, up, right) & -Q_D(s_1, left, right) & 1 \\ -Q_D(s_1, up, down) & -Q_D(s_1, left, down) & 1 \end{bmatrix}$$

The matrix b becomes $b = [0 \quad 0]^T$. The equivalency condition, which states that the action probabilities sum to 1, is given by

$$A_{eq} = [\pi_D(s_1, up) \qquad \pi_D(s_1, left)]$$

and $b_{eq} = 1$.

For the sake of completeness, we also write out the matrix equations for the invader as

$$\min_{x_I} f_I^T x_I$$

where $f_I^T = [0, 0, -1]$ and $x_I^T = [\pi_I(s_1, right), \pi_I(s_1, down), V_I(s_1)]$, subject to the constraint

$$-Q_I(s_1, right, up)\pi_I(s_1, right) - Q_I(s_1, down, up)\pi_I(s_1, down) + V_I(s_1) \le 0$$
$$-Q_I(s_1, right, left)\pi_D(s_1, right) - Q_I(s_1, down, left)\pi_I(s_1, down) + V_I(s_1) \le 0$$

and

$$\pi_I(s_1, right) + \pi_I(s_1, down) = 1$$

Then the matrix A becomes

$$A = \begin{bmatrix} -Q_I(s_1, up, right) & -Q_I(s_1, left, right) & 1 \\ -Q_I(s_1, up, down) & -Q_I(s_1, left, down) & 1 \end{bmatrix}$$

The matrix b becomes $b = [0 \quad 0]^T$. The equivalency condition, which states that the action probabilities sum to 1, is given by

$$A_{eq} = [\pi_I(s_1, right) \quad \pi_I(s_1, down)]$$

and $b_{eq} = 1$.

We first apply the minimax-Q algorithm to the game. To better examine the performance of the minimax-Q algorithm, we use the same parameter settings as in Reference 1. We use the ε-greedy policy as the exploration-exploitation strategy. The ε-greedy policy is defined such that the player chooses an action randomly from the player's action set with a probability ε and a greedy action with a probability $1 - \varepsilon$. The greedy parameter ε is given as 0.2. The learning rate α is chosen such that the value of the learning rate will decay to 0.01 after one million iterations. The discount factor γ is set to 0.9. The number of iterations represents the number of times the step 2 is repeated in Algorithm 4.1. After learning, we plot the players' learned strategies in Fig. 4-3.

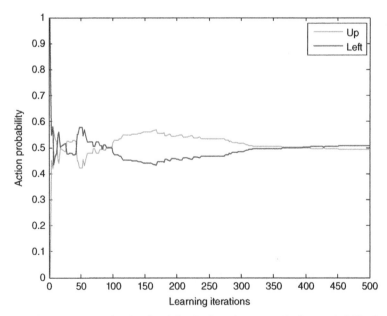

Fig. 4-3. Minimax Q-learning for the defender/invader game. Action probability for the defender.

4.4   Nash Q-Learning

The Nash Q-learning algorithm, first introduced in Reference 7, extends the minimax-Q algorithm [1] from zero-sum stochastic to general-sum stochastic games. In the Nash Q-learning algorithm, the Nash Q-values need to be calculated at each state in order to update the action-value functions and find the equilibrium strategies. Although Nash Q-learning is applied to general-sum stochastic games, the conditions for the convergence to a Nash equilibrium do not cover a correspondingly general class of environments [8]. The corresponding class of environments are actually limited to cases where the game being learned only has coordination or adversarial equilibrium [8, 10]. The Nash Q-Learning algorithm is shown in Algorithm 4.2.

Algorithm 4.2 Nash Q-learning algorithm

1: Initialize $Q_i(s, a_1, \dots, a_n) = 0$, $\forall a_i \in A_i$, $i = 1, \dots, n$
2: **for** Each iteration **do**
3:     Player i takes an action a_i from current state s based on an exploration-exploitation strategy
4:     At the subsequent state s', player i observes the rewards received from all the players $r_1, \dots, r_n$, and all the players' actions taken at the previous state s.
5:     Update $Q_i(s, a_1, \dots, a_n)$:

$$Q_i(s, a_1, \dots, a_n) \leftarrow (1 - \alpha)Q_i(s, a_1, \dots, a_n) + \alpha\left[r_i + \gamma \text{Nash}Q_i(s')\right] \quad (4.15)$$

   where α is the learning rate and γ is the discount factor
6:     Update $\text{Nash}Q_i(s)$ and $\pi_i(s)$ using quadratic programming
7: **end for**

To guarantee the convergence to Nash equilibria in general-sum stochastic games, the Nash Q-learning algorithm needs to hold the following condition during learning: every stage game (or state-specific matrix game) has a global optimal point or a saddle point for all time steps and all the states [8]. Since the above strict condition is defined in terms of the stage games as perceived during learning, it cannot be evaluated in terms of the actual game being learned [8]. Similar to the minimax Q-learning, the Nash Q-learning algorithm needs to solve the appropriate search algorithm (such as the Lemke–Howson algorithm) at each iteration in order to obtain the Nash Q-values, which leads to a slow learning process.

The Nash Q-learning algorithm extends the Q-learning algorithm to the non-cooperative multiagent context. The learning agent maintains a Q-function over joint actions and performs updates by assuming the existence of Nash equilibrium. This algorithm provably converges given certain constraints on stage games. Hu and Wellman [8] find that agents are more likely to reach a joint optimal path with Nash-Q than with single-agent Q-learning. Although the single-agent properties of Q-learning do not transfer to the multiagent case, the ease of application does.

Direct implementation of Q-learning to the multiagent context is affected by three issues. The environment is no longer stationary, the familiar guarantees are no longer true, and the nonstationary environment is populated by other agents which we assume to be rational. Explicitly accounting for the fact that the agent is operating with other "rational" agents improves the learning process [1, 8, 11]. The reward depends on the joint actions of the other learners. The default or baseline solution concept for general-sum games is the Nash equilibrium. In the framework of general-sum stochastic games, we define optimal Q-values as those values received in Nash equilibrium and refer to them as *Nash Q-values*. The goal is to find the Nash Q-values through repeated play. Agents have to learn the behavior of the other agents and then determine their own best response. In Reference 8, two grid games are proposed. In grid game 1, there are three equally valued global optimal points. Grid game 2 does not have a saddle point or a global optimal point but three sets of other Nash equilibria and in these cases the algorithm does not always converge.

As we recall from the single-agent Q-learning, the agent's objective is to find a strategy (policy) π so as to maximize the sum of discounted future rewards, given by

$$V(s, \pi) = \sum_{t=0}^{\infty} \beta' E(r_t | \pi, s_0 = s) \tag{4.16}$$

The search algorithm attempts to find the stationary point of

$$V(s, \pi^*) = \max_a \left\{ r(s, a) + \beta \sum_{s'} p(s' | s, a,) v(s', \pi^*) \right\} \tag{4.17}$$

The solution to the above Bellman equation is guaranteed to be optimal. Define the optimal Q-function as

$$Q^*(s, a) = r(s, a) + \beta \sum_{s'} p(s'|s, a)v(s', \pi^*) \tag{4.18}$$

The term $Q^*(s, a)$ is the total discounted reward for taking action a, in state s and then following the optimal policy thereafter. By definition, we have

$$V(s, \pi^*) = \max_a Q^*(s, a) \tag{4.19}$$

If we know $Q^*(s, a)$, then the problem can be solved simply by taking the action that maximizes $\max_a Q^*(s, a)$. The Q-learning algorithm is a stochastic approximation algorithm, given by

$$Q_{t+1}(s_t, a_t) = (1 - \alpha_t)Q_t(s_t, a_t) + \alpha_t(r_t + \beta \max_a Q_t(s_{t+1}, a)) \tag{4.20}$$

Hu and Wellman [8] limit their work to stationary strategies and state the following theorem:

Theorem 4.1   Theorem 4 (Fink, 1964): Every n-player discounted stochastic game possesses at least one Nash equilibrium point in stationary strategies.

The Q-learning algorithm is extended to multiagent games on the basis of the framework of stochastic games. The Nash Q-value is the expected sum of future discounted rewards when all agents follow Nash equilibrium strategies from the next step onwards. This is different from the single-agent case where the future rewards are based only on the agent's own optimal strategy.

Definition 4.1 (Hu and Wellman [8]) *Agent i's Nash Q-function is defined over $(s, a^1, \ldots, a^n)$ as the sum of agent i's current reward plus its future rewards when all agents follow a joint Nash equilibrium strategy. That is,*

$$Q^i(s, a^1, \ldots, a^n) = r^i(s, a^1, \ldots, a^n) + \beta \sum_{s' \in S} p(s'|s, a^1, \ldots, a^n)v^i(s', \pi^1, \ldots, \pi^n)$$

$$\tag{4.21}$$

where $(\pi^1, \ldots, \pi^n)$ is the joint Nash strategy, $r^i(s, a^1, \ldots, a^n)$ is agent i's reward in state s and under joint action $(a^1, \ldots, a^n)$, and $V^i(s', \pi^1, \ldots, \pi^n)$ is agent i's total discounted reward from s' given the other agents follow their Nash equilibrium strategy.

In the Nash Q-learning algorithm, the multiagent Q-learning algorithm updates with future Nash equilibrium payoffs, whereas the single-agent

Q-learning updates are based on the maximum A-values in the agent's own payoff table. To know the Nash equilibrium payoffs, the agent must also know the reward received by the other agents. The agent must be able to observe these rewards in some way. We define the difference between the Nash equilibria for a stage game and for a stochastic game.

Definition 4.2 (Hu and Wellman [8]) *An n-player stage game is defined as* $(M^1, \dots, M^n)$, *where for* $k = 1, \dots, n$ *the term* M^k *is agent* k's *payoff function over the space of joint actions* $M^k = \{r^k(a^1, \dots, a^n) | a^1 \in A^1, \dots, a^n \in A^n\}$ *and* r^k *is agent* k's *payoff.*

The Nash Q-learning algorithm executes as follows: initialize the Q-table as $Q_0^i(s, a^1, \dots, a^n) = 0, \forall s \in S, a^1 \in A^1, \dots, a^n \in A^n$. At each time t, agent i observes the current state and takes an action. Then the agent observes its reward, the rewards received by the other agents, and the new state. The agent then computes a Nash equilibrium at the new state for the stage or matrix game, $Q_t^1(s'), \dots, Q_t^n(s')$, and updates its Q-values as

$$Q_{t+1}^i(s, a^1, \dots, a^n) = (1 - \alpha_t)Q_t^i(s, a^1, \dots, a^n) + \alpha_t[r_t^i + \beta NashQ_t^i(s')] \quad (4.22)$$

The notation for all players or agents playing their Nash equilibrium strategy is $\pi_1(s'), \dots, \pi_n(s')$, and the term $Q_t^i(s', \pi_1(s'), \dots, \pi_n(s'))$ is the Nash equilibrium payoff for state s'. Therefore, for agents to determine the Nash equilibrium, they each have to know the other agents' Q-values. Thus agent i must learn the Q-values for the other agents. For example, agent i may initialize the Q-values for the other agents as $Q_0^j(s, a^1, \dots, a^n) = 0$ for all j and all $s, a^1, \dots, a^n$. Agent i observes the other agents' rewards and actions and then updates the other agents' Q-values. The update rule is the same as above, that is

$$Q_{t+1}^j(s, a^1, \dots, a^n) = (1 - \alpha_t)Q_t^j(s, a^1, \dots, a^n) + \alpha_t[r_t^j + \beta NashQ_t^j(s')] \quad (4.23)$$

Therefore, only the entry in the Q-table associated with the current state and action is updated. Although the Nash Q-learning algorithm can be written easily as in Algorithm 4.2, the algorithm is extremely complex. The user has to maintain multiple Q-tables and then compute a Nash equilibrium that all agents agree upon. One of the difficulties in implementing the Nash Q-learning algorithm is the computation of the Nash equilibrium. Hu and Wellman [8] use the Lemke–Howson algorithm [12]. We will present a detailed description of this algorithm in Section 4.6. We will present the examples of the following two grid games that are also proposed in Reference 8 and used to evaluate several other algorithms as well. The games are illustrated in Figs. 4-4 and 4-5.

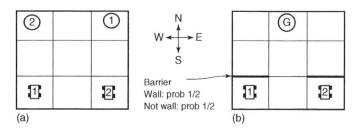

Fig. 4-4. Two stochastic games [7]. (a) Grid game 1. (b) Grid game 2.

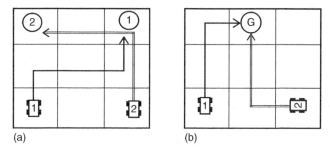

Fig. 4-5. (a) Nash equilibrium of grid game 1. (b) Nash equilibrium of grid game 2. Reproduced from [8] with permission from MIT press.

The agents can move Up, Down Right and Left. If two agents move to the same cell, then they bounce back unless it is the goal cell. The game ends when an agent reaches the goal state. If both agents reach the goal state at the same time, then both get rewarded with positive payoff. Agents do not initially know their goals or their payoffs. The agents choose actions simultaneously. The grid cells are defined as starting with cell state 0 at the bottom left corner and incrementing from left to right until the top right hand corner which is cell state 8. For example the bottom right hand corner is cell state 2. The action space is $A^i = \{Left, Right, Down, Up\}$ and a given state is given by $s = (l^1, l^2)$ where l^1 and l^2 represent the locations of the first and the second agents. If an agent reaches its goal, then it gets a reward of 100. If the agents collide, both agents bounce back to their original position and get a negative reward of -1, and if the agent moves to an empty cell, it receives 0 reward.

In grid game 1, the state transitions are deterministic, and in grid game 2 the transition through the barrier is 50%. For example, if agent 1 chooses action Up, and agent 2 chooses action Left, from state $(0, 2)$, we get the state transition probabilities as

$$P((0, 1)|(0, 2), Up, Left) = 0.5$$

and

$$P((3,1)|(0,2), Up, Left) = 0.5$$

Similarly, we have $P((1,2)|(0,2), Right, Up) = 0.5$ and $P((1,5)|(0,2), Right, Up) = 0.5$. The Nash Q-learning algorithm assumes that the strategies are stationary. A stationary strategy assigns probability distributions over an agent's actions based on the current state, regardless of the history. This means that, if the agents are in the same state, then the action strategy would be the same for all the agents.

Assuming we have pure strategies, such as in game 1, and the strategy is based only on the location of the agents, then the strategies represent a path. The notation (l^1, any) refers to agent 1 being in state l^1 and agent 2 being in *any* state. Then we get a Nash equilibrium strategy as in Table 4.4. One of the Nash equilibrium strategies is depicted in Fig. 4-5. The value of the game for agent 1 is defined so that its accumulated reward when both agents follow their Nash equilibrium is given as

$$v^1(s_0) = \sum_t \beta^t E(r_t | \pi^1, \pi^2, s_0) \tag{4.24}$$

In grid game 1 and initial state $s_0 = (0, 2)$, this becomes, given $\beta = 0.99$

$$v^1(s_0) = 0 + 0.99 \times 0 + 0.99 * 2 \times 0 + 0.99^3 \times 100 = 97 \tag{4.25}$$

Based on the value for each state, we can derive the Nash Q-values for agent 1 in state s_0 as

$$Q^1(s_0, a^1, a^2) = r^1(s_0, a^1, a^2) + \beta \sum_{s'} p(s'|s_0, a^1, a^2) v^1(s') \tag{4.26}$$

We will evaluate the Q-value for different actions for agent 1 in state $(0, 2)$. Let us start with the action (Right, Left). If we take the action (Right, Left), then the two agents will bump into each other. This will give a penalty of $r^1((0,2), Right, Left) = -1$. Therefore, the Q-value for taking the action (Right,

Table 4.4   States and strategies.

State	$\pi^1(s)$
(0, 2)	Up
(3, 5)	Right
(4, 8)	Right
(5, any)	Up

Left) in state (0, 2) and then following the optimal path afterwards, from state (0, 2) (because the agents bounce back to state (0, 2)) is

$$Q^1(s_0, Right, Left) = -1 + \beta v^1((0,2)) \tag{4.27}$$

We have already computed that $v^1((0,2)) = 97$, and therefore

$$Q^1(s_0, Right, Left) = -1 + 0.99 \times 97 = 95.1 \tag{4.28}$$

But, for the Q-value for both agents going Up, we get

$$Q^1(s_0, Up, Up) = 0 + 0.99v^1(3,5) = 97 \tag{4.29}$$

The Nash Q-values for the agents when in state $(0, 2)$ are given in Table 4.5. For grid game 2, we can derive the Nash Q-values. In this case, it is more complicated because we do not have deterministic state transitions. Let us start the agents in state $(0, 1)$ as shown in Fig. 4-6. Then the optimal path for this position is one in which agent 2 takes two steps Up and gets the reward. Agent 1 cannot get to the goal before agent 2. Therefore, the value for agent 1 is

$$v^1(0,1) = 0 + 0.99 \times 0 + 0.99^2 \times 0 = 0 \tag{4.30}$$

Table 4.5 Grid game 1: Nash Q-values in state (0, 2).

Action	Left	Up
Right	95.1, 95.1	97, 97
Up	97, 97	97, 97

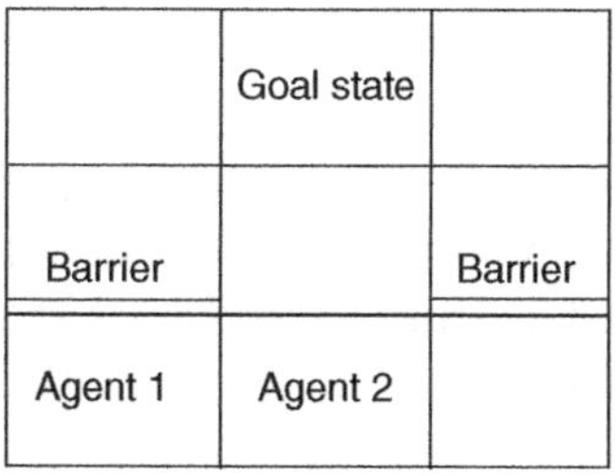

	Goal state	
Barrier		Barrier
Agent 1	Agent 2	

Fig. 4-6. Grid game with barriers, start position (0,1).

However, if we start in state $(1, 2)$, then agent 1 takes two steps Up and wins, and the value for agent 1 is

$$v^1(1, 2) = 0 + 0.99 \times 100 = 99 \tag{4.31}$$

However, if we start in state $(0, 2)$, then we can only compute the value in expectation because if the agents choose the action Up, there is only 50% probability that they would go Up and a 50% probability that they stay in the same spot. For starting in state $(0, 2)$, we get the Q-value as

$$Q^1((0, 2), Right, Left) = -1 + 0.99v^1((0, 2)) \tag{4.32}$$

The next action is $Q^1((0, 2), Right, Up)$. In this case, there is only a 50% chance that the agent goes Up. So we write out the Q^1-value as

$$Q^1(s_0, Right, Up) = 0 + 0.99 \left(\frac{1}{2}v^1(1, 2) + \frac{1}{2}v^1(1, 5) \right) \tag{4.33}$$

If the agents take actions (Right, Up), then we may end up in the position $(1, 2)$ or $(1, 5)$. Now, recall that if the agents are in state $(1, 2)$, then the optimal solution from there is to take two steps Up. Similarly, the agent in state $(1, 5)$ has the optimal solution of two steps, Up and Left. Recall, the values of $v^1(1, 2) = v^1(1, 5) = 0 + 0.99 \times 100 = 99$. Then we can compute

$$Q^1((0, 2), Right, Up) = 0 + 0.99 \left(\frac{1}{2}(0.99) + \frac{1}{2}(0.99) \right) = 0.98 \tag{4.34}$$

Now let us take the case of $Q^1((0, 2), Up, Left)$. In this case, there is only 50% chance that agent 1 moves Up and 50% chance that the agent stays in the same position. Then we can compute

$$Q^1((0, 2), Up, Left) = 0 + 0.99 \left(\frac{1}{2}v^1(0, 1) + \frac{1}{2}v^1(3, 1) \right) \tag{4.35}$$

We already know that the value of $v^1 = 0$; this is because agent 2 will take two steps Up and win. If agent 1 gets through the barrier, then it is in position $(3, 1)$ and agent 1 then takes two steps, Up and Left, and wins at the same time as agent 2 takes two steps, Up and Up, and wins. Therefore, we get $v^1(1, 3) = 0 + 0.99 \times 100 = 0.99$, and we can compute the optimal Q-value as

$$Q^1((0, 2), Up, Left) = 0 + 0.99 \left(\frac{1}{2} \times 0 + \frac{1}{2} \times 99 \right) = 49 \tag{4.36}$$

Finally, we compute the value for $Q^1((0, 2)Up, Up)$. In this case, there are four possible outcomes for the next state. Each agent has a 50% probability of moving Up, therefore there is 25% probability for both agents to move Up at the same time. The probability of achieving state (3, 2) is 25%, state (3, 5) is 25%, state (0, 2) is 25%, and state (0, 5) is 25%. So the Q-value for $Q^1((0, 2), Up, Up)$ is

$$Q^1((0, 2), Up, Up) = 0 + 0.99 \left(\frac{1}{4} v^1(3, 2) + \frac{1}{4} v^1(3, 5) + \frac{1}{4} v^1(0.2) + \frac{1}{4} v^1(0, 5) \right)$$
(4.37)

We know the value of $v^1(3, 2) = 99$, $v^1(3, 5) = 99$, and $v^1(0, 5) = 0$. Then,

$$Q^1((0, 2), Up, Up) = 0 + 0.99 \left(\frac{1}{4} v^1(0, 2) + \frac{1}{4} \times 99 + \frac{1}{4} \times 99 + \frac{1}{4} \times 0 \right)$$

$$= 0.99 \times \frac{1}{4} v^1(0, 2) + 24.5 + 24.5$$

$$= 0.99 \times \frac{1}{4} v^1(0, 2) + 49$$

Now let us define $R_1 = v^1(0, 2)$ to be agent 1's optimal value by following a Nash equilibrium strategy starting from state (0, 2). The obvious Nash equilibrium for agent 1 of game 2 from the initial state (0, 2) is (Right, Up). Given that (Right, Up) is the Nash equilibrium then, $v^1(0, 2) = 0.98$. Then we can derive the other values in the Q-table. On the other hand, the obvious Nash equilibrium for the second agent is (Up, Left), but for agent 1 $Q^1((0, 2), Up, Left) = 49$. Furthermore, there is also a mixed strategy of $(\{P(Right) = 0.97, P(Up) = 0.03\}, \{P(Left) = 0.97, and\ P(Up) = 0.03\})$. There are three sets of possible Nash equilibria for grid game 2.

4.4.1   The Learning Process

Let us say that learning of agent 1 begins by initializing the Q-table as $Q^1(s, a^1, a^2) = 0$ for all s, a^1, and a^2. These are agent i's internal beliefs, and have nothing to do with the other agent. We start the game from the initial state (0, 2). The agents then move simultaneously and observe the action taken by the agents and the rewards that the agents receive. The agents then update their Q-tables according to the following:

$$Q^j_{i+1}(s, a^1, a^2) = (1 - \alpha_t)Q^j_i(s, a^1, a^2) + \alpha_t[r^j_t + \gamma NashQ^j_t(s)]$$ (4.38)

The process is repeated in the next state until the goal state is reached. Then a new game starts and each agent is randomly assigned a new starting position except for the goal state. The training stops after 5000 episodes. Each episode takes approximately eight steps. Therefore, one experiment takes 40,000 steps. Furthermore, the learning rate is given by

$$\alpha_t(s, a^1, a^2) = \frac{1}{\eta_t(s, a^1, a^2)} \tag{4.39}$$

where $\eta_t(s, a^1, a^2)$ is the number of times the states and actions (s, a^1, a^2) have been visited.

If we look at the algorithm, we see that we have to compute the Nash equilibrium of the stage game with $(Q^1(s^1) \text{ and } Q^2(s^1))$. There may be a choice between multiple Nash equilibria. Hu and Wellman [8] use the Lemke–Howson [12] algorithm to compute the Nash equilibrium for the two-player game. The Lemke–Howson algorithm resembles the simplex algorithm from linear programming.

4.5 The Simplex Algorithm

The simplex algorithm is a well-known algorithm for solving linear programming problems. These are problems of maximizing a utility function or a cost function subject to a number of linear constraints. The linear programming model is as follows: Maximize

$$V = \sum_{j=1}^{n} c_j x_j \tag{4.40}$$

subject to the set of constraints

$$\sum_{j=1}^{n} a_{ij} x_j \le b_i \quad i = 1, 2, \dots, m \text{ and } x_j \ge 0 \tag{4.41}$$

The simplex algorithm searches for vertices of a polytope that defines the accessible region of the solution space. One goes from vertex to vertex to find the solution. The algorithm assumes nonnegativity. Inequalities are converted into equalities by adding slack variables. Let us take the following example. This example is taken from *Design and Planning of Engineering Systems* [13] (Fig. 4-7). Maximize

$$V = 13x_1 + 11x_2 \tag{4.42}$$

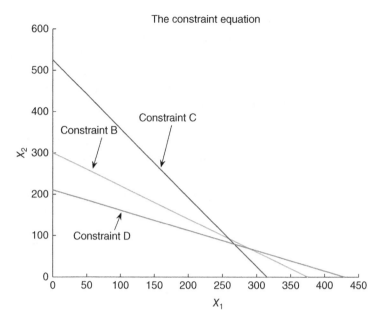

Fig. 4-7. Constraint equations plotted for the simplex method.

subject to the constraints

$$4x_1 + 5x_2 \leq 1500$$
$$5x_1 + 3x_2 \leq 1575$$
$$x_1 + 2x_2 \leq 420 \qquad (4.43)$$

and the nonnegativity constraint $x_1 \geq 0$ and $x_2 \geq 0$. For each one of the constraints, we add a slack variable x_3, x_4, or x_5. We therefore convert the inequality constraints into equality constraints, as

$$4x_1 + 5x_2 + x_3 = 1500$$
$$5x_1 + 3x_2 + x_4 = 1575$$
$$x_1 + 2x_2 + x_5 = 420 \qquad (4.44)$$

The slack variables represent how much room one has before one hits the constraint. One initializes the system to start with the variables set to $x_1 = 0$ and $x_2 = 0$. This gives the solution to the slack variables as $x^T = [0, 0, 0, 1500, 1575, 420]$. The nonzero variables are known as the *variables in the basis*. In this first step we have selected, the origin as the first vertex and the value of the reward or value function z is zero, $z = 0$. We now have to move to the next extremal point. This is equivalent to moving one variable out of the basis and moving another variable into the basis. The next step is to determine which variable leaves the basis and which one enters the basis. We put into the basis the variable that causes the greatest growth in the payoff or reward function. Therefore, in this case we put x_1 into the basis because the payoff function will grow by 13 units for every unit of x_1 whereas it only grows by 11 units for every unit of x_2. Then the first constraint is reached when $x_1 = 1500/4 = 375$ and then $x_3 = 0$. The second constraint is reached when $x_1 = 1575/5 = 315$ and then $x_4 = 0$. The final constraint is reached when $x_1 = 420$ and $x_5 = 0$. Therefore, x_1 hits constraint C first when it is equal to 315. Then x_1 enters the basis, and $x_4 = 0$ leaves the basis. We then write the variables in the basis in terms of only the variables not in the basis. This is a form of Gaussian elimination. At the next vertex we get $x_1 = 315$, $x_2 = 0$, and $x_4 = 0$. We then write the system of equations as

$$z - 13x_1 - 11x_2 = 0 \qquad \text{A}$$

$$4x_1 + 5x_2 + x_3 \leq 1500 \qquad \text{B}$$

$$5x_1 + 3x_2 + x_4 \leq 1575 \qquad \text{C}$$

$$x_1 + 2x_2 + x_5 \leq 420 \qquad \text{D} \qquad (4.45)$$

Recall that it is the x_4 slack variable that goes to 0. Therefore, we use Eq. C, to solve for x_1. Essentially, using Gaussian elimination we multiply Eq. C by 13/5 and add Eq. C to Eq. A and get

$$z - 16/5x_2 + 13/5x_4 = 4095 \quad \text{A1}$$

Then multiply Eq. C by $-4/5$ and add to Eq. B and we get

$$13/5x_2 + x_3 - 4/5x_4 = 240 \quad \text{B1}$$

Similarly, divide Eq. C by 5 and we get

$$x_1 + 3/5x_2 + 1/5x_4 = 315 \quad \text{C1}$$

And, finally, multiply Eq. C by $-1/5$ and add it to Eq. D and we get

$$7/5x_2 - 1/5x_4 + x_5 = 105 \quad \text{D1}$$

We see from these equations that, if x_2 increases, then so does the payoff function z. This is because the coefficient of x_2 in Eq. A1 is negative. So, x_2 will enter the basis and either x_3 or x_5 will leave the basis. The slack variable x_4 remains zero because we are moving along the edge defined by constraint C. Along this edge, the slack variable is zero.

Therefore, when x_3 and x_4 are zero, x_2 can grow to $5.24/13 = 92.3$; if x_4 and x_5 are zero, then x_2 can grow to $5105/7 = 75$; and if x_1 and x_4 are zero, then x_2 can grow to $5315/3 = 525$. The limiting case is when x_5 is zero and $x_4 = 0$. This represents the intersection of the two constraints C and D. Then we use Gaussian elimination again and write out the equations in terms of x_4 and x_5 only. Then we multiply Eq. D1 by $16/5 \times 5/7$ and add it to Eq. A1 and get

$$z + 15/7x_4 + 16/7x_5 = 4335 \quad \text{A2}$$

We can longer increase z without violating a constraint condition. Similarly, we remove x_2 from Eq. B1 by multiplying Eq. D1 by $-13/5 \times 5/7$ and add it to Eq. B1. This gives

$$x_3 - 3/7x_4 - 13/7x_5 = 45 \quad \text{B2}$$

Similarly, for Eq. C1 we multiply Eq. D1 by $-3/5 \times 5/7$ and add it to Eq. C1 and get

$$x_1 + 10/35x_4 - 3/7x_5 = 270 \quad \text{C2}$$

Finally, we multiply Eq. D1 by $5/7$ and get

$$x_2 - 1/7x_4 + 5/7x_5 = 75 \quad \text{D2}$$

To summarize, we can write the equations as

$$z + 15/7x_4 + 16/7x_5 = 4335 \quad \text{A2}$$

$$x_3 - 3/7x_4 - 13/7x_5 = 45 \quad \text{B2}$$

$$x_1 + 10/35x_4 - 3/7x_5 = 270 \quad \text{C2}$$

$$x_2 - 1/7x_4 + 5/7x_5 = 75 \quad \text{D2} \tag{4.46}$$

Recall that the slack variables x_4 and x_5 are zero at the intersection of the constraint equations C and D. Then the optimal value for the cost or payoff function is $z = 4335$ and $x_1 = 270$ and $x_2 = 75$ and the slack value of the constraint given by Eq. B is $x_3 = 45$.

4.6 The Lemke–Howson Algorithm

This algorithm is applicable to the two-player game. Player 1 has m actions and player 2 has n actions. We label the actions of player 1 as $M = \{1, \dots, m\}$ and the actions of player 2 as $N = \{m + 1, \dots, m + n\}$. The payoff for the players is given by the usual payoff matrix. We have two payoff matrices, matrix A and matrix B for player 1 and player 2, respectively. If there is a mixed strategy, given by the action profile (x, y), then the payoff for player 1 is $x^T A y$ and the payoff for player 2 is $x^T B y$. The support of a vector is defined as $supp\{\cdot\}$ and represents the indices of elements of the vector that are nonzero. The term x denotes all possible mixed strategies for agent 1, and y denotes all possible mixed strategies for agent 2. We assume that A and B have all positive elements and that neither A nor B have all zero rows or columns. Let B_j denote the column of B corresponding to action j, and let A^i denote the row of A corresponding to action i. Define the two polytopes

$$P_1 = \{x \in R^m | (\forall i \in M : x_i \geq 0) \qquad (\forall j \in N : x^T B_j \leq 1)\} \qquad (4.47)$$

and

$$P_2 = \{y \in R^n | (\forall j \in N : y_j \geq 0) \qquad (\forall i \in M : A^i y \leq 1)\} \qquad (4.48)$$

The strategy is given by $nrml(x) := (\sum_i x_i)^{-1} x$. The inequalities that define P_1 have the following meaning:

- if $x \in P_1$ meets $x_i \geq 0$ with equality, for example, $x_i = 0$, then i is not in the support of x;
- if $x \in P_1$ meets $x^T B_j \leq 1$ with equality, then j is the best response to $nrml(x)$.

The solution techniques to find the Nash equilibrium of general-sum games are the *Lemke–Howson algorithm* and its extensions. For completeness, we will present the Lemke–Howson algorithm. Furthermore, these algorithms are based on methods from linear programming such as the simplex method. The Lemke–Howson algorithm is the central algorithm of the Nash Q-learning algorithm even though the seminal paper by Hu and Wellman [8] only cite the algorithm in passing.

The Lemke–Howson algorithm resembles the simplex algorithm. This algorithm is used for two-player bimatrix games. The Nash Q-learning algorithm uses this algorithm on each iteration to solve for the Nash equilibrium of the two-player game. In particular, Hu and Wellman [8] use this algorithm for their examples. Let player 1 have m actions given by the set $M = \{1, \ldots, m\}$, and player 2 n actions given by the set $N = \{m+1, \ldots, m+n\}$. We represent the payoff matrix for the players as an $m \times n$ matrix. We define the payoff matrix for player 1 with the matrix A and that for player 2 as the matrix B. We think of player 1 as choosing actions that are represented by rows, and player 2 as choosing the columns. We define an m-dimensional row vector x that represents the probabilities of player 1 choosing each of the possible actions. The elements of the row vector x will sum to 1. Similarly, we define an n-dimensional column vector to represent the probabilities of player 2 choosing each action. Therefore, $x \in R^m$ is the mixed strategy for player 1, and $y \in R^n$ is a column vector representing the mixed strategy for player 2. The expected reward for player 1 can then be written as

$$R_1 = x^T A y \tag{4.49}$$

and the expected reward for player 2 can be written as

$$R_2 = x^T B y \tag{4.50}$$

Similar to the simplex method, we define the support of any strategy as those actions that do not have zero entries in x or y. We assume that matrix A has no all-zero columns and that B has no all-zero rows and that the entries of A and B are all nonnegative. Let B_j denote a column of b, and a^i denote a row of A. Then we define the two polytopes as follows:

$$P_1 = \{x \in R^m | (\forall i \in M : x_i \geq 0) \quad (\forall j \in N : x^T B_j \leq 1)\} \tag{4.51}$$

$$P_2 = \{y \in R^n | (\forall j \in N : y_i \geq 0) \quad (\forall i \in M : A^i y \leq 1)\} \tag{4.52}$$

The inequalities described in the polytopes defined above have the following meanings: If $x_i = 0$, then x_i is not in the support or basis of x. Recall this language from the description in the simplex method. The second equality constraint is $x^T B_j = 1$. This means that player 2's action j is the best response to strategy x of player 1. If the sum of the elements in column B^j is greater than 1, then the sum of the elements of x will be less than 1, and to get the strategy we will need to normalize x as

$$nrml(x) := \left(\sum_i x_i \right)^{-1} x \tag{4.53}$$

Let us define labels as follows: A strategy $x \in P_1$ has label $k \in M \cup N = \{1, 2, \ldots, m+n\}$ if either $k \in M$ and $x_k = 0$ or $k \in N$ and $x^T B_k = 1$. We then can state the following theorem:

Theorem 4.2 Suppose that $x \in P_1$ and $y \in P_2$ and neither x nor y is the all-zero vector. Then, x and y together have all labels from 1 to k, iff $(nrml(x), nrml(y))$ is a Nash equilibrium.

We give a simple example to illustrate how the Lemke–Howson algorithm is implemented. Let us take the case of a two-player two-action game, where the reward matrices have the following values:

$$A = \begin{bmatrix} 4 & 6 \\ 5 & 3 \end{bmatrix} \quad B = \begin{bmatrix} 3 & 2 \\ 1 & 4 \end{bmatrix} \tag{4.54}$$

The polytopes P_1 and P_2 are defined by the limits of the constraints $A^i y \le 1$ and $B^{jT} x \le 1$, and both $x_i \ge 0$ and $y_j \ge 0$ as illustrated in Figs. 4-8 and 4-9. We will convert these inequalities into equality constrains by adding slack variables as we did in the simplex method. Define the slack variable r_i as

$$A^i y + r_i = 1 \tag{4.55}$$

Similarly

$$B^{jT} x + s_j = 1 \tag{4.56}$$

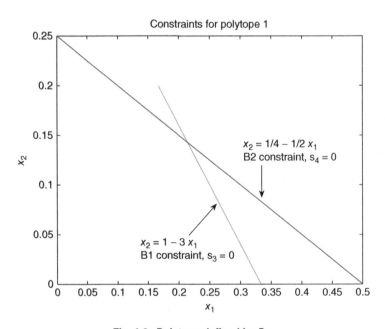

Fig. 4-8. Polytope defined by P_1.

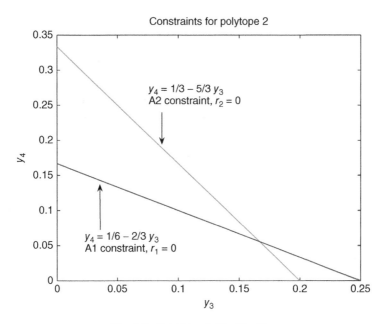

Fig. 4-9. Polytope defined by P_2.

Then, similar to the simplex method, we write $r_i = 1 - A^i y$ as

$$r_1 = 1 - 4y_3 - 6y_4 \quad \text{A1}$$
$$r_2 = 1 - 5y_3 - 3y_4 \quad \text{A2}$$

Similarly, the s constraint $s_j = 1 - B^{jT} x$ becomes

$$s_3 = 1 - 3x_1 - x_2 \quad \text{B1}$$
$$s_4 = 1 - 2x_1 - 4x_2 \quad \text{B2}$$

We can now draw out the constraint conditions as illustrated in Fig. 4-8. Taking the slack variables as being equal to zero, $s_1 = s_2 = 0$. The equality constraints for the (x_1, x_2) polytope become

$$x_2 = 1 - 3x_1$$
$$x_2 = \frac{1}{4} - \frac{1}{2}x_1$$

Similarly, for the (y_3, y_4) polytope we have the constraints

$$y_4 = \frac{1}{6} - \frac{2}{3}y_3$$

$$y_4 = \frac{1}{3} - \frac{5}{3}y_3$$

We will start by arbitrarily allowing x_1 into the basis. Then the constraint B1 limits the growth of x_1 to $1/3$, as seen in Figs. 4-8. Along the B1 constraint, the slack variable s_3 is zero. Then, from the complementary condition $s_3 y_3 = 0$, we bring y_3 into the basis. From Eq. B1 we solve for x_1 as

$$3x_1 = 1 - x_2 - s_3$$

$$x_1 = \frac{1}{3} - \frac{1}{3}x_2 - \frac{1}{3}s_3$$

where $x_2 = 0$ and $s_3 = 0$. Substituting for x_1 into constraint Eq. B2, we get

$$s_4 = 1 - 2\left(\frac{1}{3} - \frac{1}{3}x_2 - \frac{1}{3}s_3\right) - 4x_2$$

$$s_4 = 1 - \frac{2}{3} + \frac{2}{3}x_2 + \frac{2}{3}s_3 - 4x_2$$

$$s_4 = \frac{1}{3} - \frac{10}{3}x_2 + \frac{2}{3}s_3 \quad \text{B2}'$$

This is what just happened: we arbitrarily let agent 1 choose action 1 and found that condition B1 was satisfied first when $s_3 = 0$, at $x_1 = \frac{1}{3}$ and $x_2 = 0$. This means that the best response by agent 2 to agent 1 choosing action 1 is for agent 2 to take action 3, which represents the first column of the payoff function given by matrix B. This is obvious because, if agent 1 picks action 1, then agent 2 should pick action 3, which is the first column of the B matrix and then agent 2 receives a reward of 3. If agent 2 chooses action 4, then its reward would be only 2. Given that agent 1 chooses action 1, this means that agent 2 chooses action 3, and this is equivalent to having y_3 enter the basis. The limiting condition on y_3 is given by the constraint A2. We then solve for y_3 because of constraint A2 as

$$y_3 = \frac{1}{5} - \frac{3}{5}y_4 - \frac{1}{5}r_2 \quad \text{where} \quad y_4 = 0 \quad \text{and} \quad r_2 = 0$$

We then compute r_1 as

$$r_1 = 1 - 4\left(\frac{1}{5} - \frac{3}{5}y_4 - \frac{1}{5}r_2\right) - 6y_4$$

$$r_1 = \frac{1}{5} - \frac{18}{5}y_4 + \frac{4}{5}r_2 \quad \text{A2}'$$

Therefore, the best response to agent 2 taking action 3 is for agent 1 to take action 2 because $r_2 = 0$. We then pivot along the $s_3 = 0$ line in the (x_1, x_2)

polytope. We then add x_2 to the basis because we have $x_2 r_2 = 0$, and solve for x_2 from Eq. B2' as

$$\frac{10}{3} x_2 = \frac{1}{3} + \frac{2}{3} s_3 - s_4$$

$$x_2 = \frac{1}{10} + \frac{1}{5} s_3 - \frac{3}{10} s_4 \quad \text{where} \quad s_3 = s_4 = 0$$

$$x_1 = \frac{1}{3} - \frac{1}{3} \left(\frac{1}{10} + \frac{1}{5} s_3 - \frac{3}{10} s_4 \right) - \frac{1}{3} s_3$$

$$= \frac{9}{30} - \frac{16}{15} s_3 + \frac{1}{16} s_4$$

Then we choose to take y_4 into the basis because $s_4 = 0$, and it is clear that the best response to agent 1 taking action 2 is for agent 2 to take action 4 and get a reward of 4. We then solve for y_4 from Eq. A2' and get

$$\frac{18}{5} y_4 = \frac{1}{5} - r_1 + \frac{4}{5} r_2$$

$$y_4 = \frac{1}{18} - \frac{5}{18} r_1 + \frac{4}{18} r_2 \quad A''$$

Then $r_1 = 0$, and x_1 would enter the basis, but it is already in the basis so the algorithm ends. We now normalize the game strategy. The current solution is given by $x_1 = \frac{9}{30} = \frac{3}{10}, x_2 = \frac{1}{10}, y_3 = \frac{1}{6}$, and $y_4 = \frac{1}{18}$. We then normalize the strategy as

$$x_{10} = \frac{\frac{3}{10}}{\frac{3}{10} + \frac{1}{10}} = \frac{3}{4} \quad \text{and} \quad x_{20} = \frac{\frac{1}{10}}{\frac{3}{10} + \frac{1}{10}} = \frac{1}{4}$$

$$y_{30} = \frac{\frac{3}{18}}{\frac{3}{18} + \frac{1}{18}} = \frac{3}{4} \quad \text{and} \quad y_{40} = \frac{\frac{1}{18}}{\frac{3}{18} + \frac{1}{18}} = \frac{1}{4}$$

Now let us check if this works. Given that $x_0 = [\frac{3}{4}, \frac{1}{4}]^T$ and $y_0 = [\frac{3}{4}, \frac{1}{4}]^T$, the payoff to agent 1 is

$$R_1 = x_0^T A y_0$$

$$= \begin{bmatrix} \frac{3}{4} \\ \frac{1}{4} \end{bmatrix} \begin{bmatrix} 4 & 6 \\ 5 & 3 \end{bmatrix} \begin{bmatrix} \frac{3}{4} \\ \frac{1}{4} \end{bmatrix} = 4.5$$

We can also write this in the form before normalization, in which case we get the constraint condition as

$$Ay^* = \begin{bmatrix} 4 & 6 \\ 5 & 3 \end{bmatrix} \begin{bmatrix} \frac{1}{6} \\ \frac{1}{18} \end{bmatrix} = \begin{bmatrix} 1 \\ 1 \end{bmatrix}$$

which satisfies the constraint condition. We can also check the payoff for agent 2.

$$R_2 = x_0^T B y_0$$

$$= \begin{bmatrix} \frac{3}{4} \\ \frac{1}{4} \end{bmatrix} \begin{bmatrix} 3 & 2 \\ 1 & 4 \end{bmatrix} \begin{bmatrix} \frac{3}{4} \\ \frac{1}{4} \end{bmatrix} = 2.5$$

and once again we can verify the constraint conditions as

$$B^T x^* = \begin{bmatrix} 3 & 1 \\ 2 & 4 \end{bmatrix} \begin{bmatrix} \frac{3}{10} \\ \frac{1}{10} \end{bmatrix} = \begin{bmatrix} 1 \\ 1 \end{bmatrix}$$

Recall the conditions $x_0^T A y_0 \geq x^T A y_0$ and $x_0^T B y_0 \geq x_0^T B y$. This says that if agent 2 plays its best policy, and if agent 1 plays any policy, it does no better than the optimal policy x_0, and similarly for agent 2. Therefore, we now show that we cannot find any x that will make R_1 greater than the optimal policy selection. We know that the optimal payoff for agent 1 is given by $R_1^* = x_0^T A y_0 = 4.5$. Can we find a policy $x \neq x_0$ such that $R_1 = x^T A y_0 > 4.5$? We have

$$R_1 = \begin{bmatrix} x_1 \\ x_2 \end{bmatrix} \begin{bmatrix} 4 & 6 \\ 5 & 3 \end{bmatrix} \begin{bmatrix} \frac{3}{4} \\ \frac{1}{4} \end{bmatrix}$$

$$= \begin{bmatrix} x_1 \\ x_2 \end{bmatrix}^T \begin{bmatrix} 4.5 \\ 4.5 \end{bmatrix} \quad \forall \quad x_1 \quad \text{and}$$

$$x_2 \text{ s.t.} \quad x_1 + x_2 = 1 \tag{4.57}$$

Similarly, for agent 2 we have $R_2 = x_0^T B y_0 \geq x_0^T B y$, but we know that $x_0^T B = [2.5, 2.5]$ and that $x^T B = [1, 1]$. Therefore, if agent 2 picks its optimal policy y_0, then agent 1 will always get the payoff 4.5, regardless of what strategy it uses; similarly, if agent 1 picks its optimal strategy x_0, then agent 2 gets its optimal payoff off 2.5 regardless of what strategy it picks. We now see the importance of setting the constraint conditions for agent 1 to ensure that $B^T x = 1$ and for agent 2 $Ay = 1$. Then for condition $B^T x = 1$, this means that the payoff to agent 2 is limited by the strategy of agent 1, and similarly the condition $Ay = 1$ means that the payoff to agent 1 is limited by the strategy of agent 2, yielding the maximum payoff for each constrained by the other.

4.7 Nash-Q Implementation

Our goal with the Nash-Q algorithm is to reproduce the results obtained by Hu and Wellman. The results were very conclusive in grid games 1 and 2. We wanted to recreate the situation when both agents were Nash-Q learners. Hu and Wellman's observations showed that both agents learned a Nash equilibrium strategy 100% of the time. The high percentage is a very good indicator that the algorithm works very well in that particular situation.

The Lemke–Howson algorithm was used to find the Nash equilibrium of bimatrix games. We used the original version from Hu and Wellman and adapted it for our environment in MATLAB. The algorithm is designed to find more than one Nash equilibrium when they exist. It is based on the work from Reference 14. The inputs are the two payoff matrices for both players, and the outputs are the Nash equilibria. In our multiple implementations, we kept the parameters from Reference 8. The author had their agents play for 5000 games. The number of games necessary to reach convergence is affected by the variation of the learning rate α. Where α is inversely proportional to the number of times each state tuples (s, a^1, a^2) are visited. We considered 5000 a reasonable number of games because after some testing we calculated that each state would be visited on average 93 times.

We started the implementation with a more general approach to the learning technique than in Reference 8. In our first implementation, our agents were always able to choose any of the four actions. This meant that if the agent chooses to go into a wall, it would be bounced back and would receive either a negative reward or no reward. Also, our Lemke–Howson algorithm would use the Nash Q-values of all four actions for each state. It meant that the algorithm was calculating more than it needed. The results obtained were not conclusive. In our second implementation, we changed the code so that the agents would not be able to choose an action that would lead out of

Table 4.6   Grid game 1.
Nash Q-values in state
(1,3).

	Up	Left
Up	96,95	92,92
Right	85,85	89,85

bounds. The fact that Nash-Q is an offline learner helped us to create the boundaries when choosing randomly the next step. This change positively affected the results, and we got a success rate of about 25%. This was still far from the success rate noted in the literature. In our last implementation, the Lemke–Howson algorithm was used only on the possible actions. This means that the agent would calculate the Nash equilibrium strategy of the next state with only the actions that are allowed in that state. We were able to achieve 100% success in getting a Nash equilibrium strategy for two Nash-Q learners. Table 4.6 shows the value of the Nash Q-values in the starting state (1,3).

We decided to use the same method of Hu and Wellman to confirm their results. This means that the agent will choose a random action in every state. The starting positions of the players change in every game. They start in a random position except for their goal cell. This ensures that each state is visited often enough. According to Reference 8, the learning rate depends on the number of times each state-action tuple has been visited. The value of α is $\alpha(s, a^1, a^2) = \frac{1}{n_t(s,a^1,a^2)}$, where n_t is the number of times the game was in the state-action tuple (s, a^1, a^2) [8]. This allows the learning rate to decay until the state-action tuple is visited often enough. We found that if we remove the states where both players occupied the same cell, the states where one or both of the players are in their goal cell, and the inaccessible actions (the players cannot try to move into the wall), we would get 424 different state-action tuples of the form (s, a^1, a^2) [8]. The learning rate would be negligible after 500 visits with a value of $\alpha = 0.002$.

We also implemented an online version of the algorithm. In this version, the agents start each episode from the state $s(1, 3)$, which is the original starting position. The main difference is that the agent now uses an exploit-explore learning technique where the agent choose a random action with a probability of $1 - \varepsilon$ and the Nash equilibrium strategy with a probability of ε. The value of ε varies during learning as $\varepsilon(s) = \frac{1}{n_t(s)}$, where n_t is the number of times the game was in the state s [8]. This means that the chance that the agent will choose a random action increases with time. This is not desirable for an online learning agent, because it would cause the average reward to decrease with time. It is necessary for the Nash-Q agent to visit as many different states as possible to ensure convergence to a Nash equilibrium strategy.

Our final results correspond perfectly with those in the literature when looking at two Nash-Q learners playing grid games 1 and 2. We tested each grid game 20 times, and the agents found the Nash equilibrium strategy 100% of the time. We also kept track of the performance of each agent. The performance was calculated by the average reward per step the agent was able to accumulate. The performance of an agent is important because it tells us how well the agent can optimize their policy. To ensure that the values reflect not only the results of one game, we took the average of the value over five games. It also gave us a smoother curve on the graphs which made it easier to read. In Figs. 4-10–4-12, we illustrate the performance of the algorithm in grid game 1 when both agents are Nash-Q learners. It shows the average rewards per step

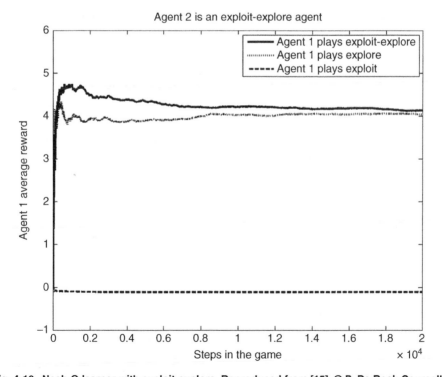

Fig. 4-10. Nash-Q learner with exploit-explore. Reproduced from [15], © P. De Beck-Courcelle.

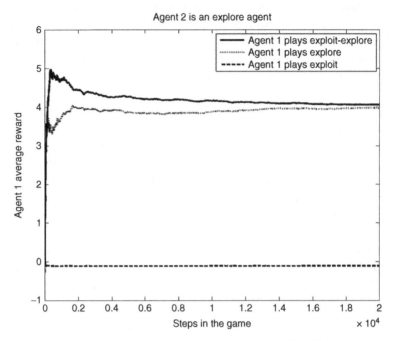

Fig. 4-11. Nash-Q learner with explore only. Reproduced from [15], © P. De Beck-Courcelle.

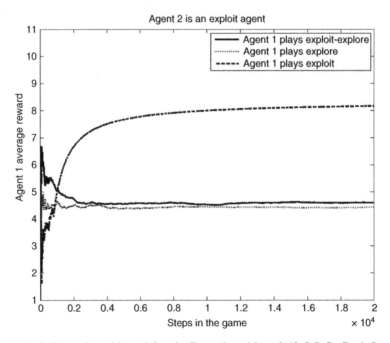

Fig. 4-12. Nash-Q learning with exploit only. Reproduced from [15], © P. De Beck-Courcelle.

when the two agents use the three learning technique: exploit-explore, explore only, and exploit only.

The Nash-Q algorithm converges to a Nash equilibrium strategy only when all the assumptions and conditions are met. The algorithm is also very demanding on computation time. Its needs to keep track of all the players' actions and rewards in a separate Q-table. It also needs to evaluate the Nash equilibrium at every step to update its Q-values. The Lemke–Howson algorithm is equivalent to the simplex algorithm in terms of computation. Both these actions require a large amount of processing power and consume a lot of time.

4.8 Friend-or-Foe Q-Learning

For a two-player zero-sum stochastic game, the minimax-Q algorithm [1] is well suited for the players to learn a Nash equilibrium in the game. For general-sum stochastic games, Littman proposed a friend-or-foe Q-learning (FFQ) algorithm such that a learner is told to treat the other players as either a "friend" or a "foe" [10]. The FFQ algorithm assumes that the players in a general-sum stochastic game can be grouped into two types: player i's friends and player i's foes. Player i's friends are assumed to work together to maximize player i's value, while player i's foes work together to minimize the value [10]. Thus, an n-player general-sum stochastic game can be treated as a two-player zero-sum game with an extended action set [10].

The FFQ algorithm for player i is given in Algorithm 4.3. Note that the FFQ algorithm is different from the minimax-Q algorithm for a two-team zero-sum stochastic game. In a two-team zero-sum stochastic game, a team leader controls the team players' actions and maintains the value of the state for the whole team. The received reward is also the whole team's reward. For the FFQ algorithm, there is no team leader to send commands to control the team players' actions. The FFQ player chooses its own action and maintains its own state-value function and equilibrium strategy. In order to update the action-value function $Q_i(s, a_1, \ldots, a_{n_1}, o_1, \ldots, o_{n_2})$, the FFQ player needs to observe its friends and opponents' actions at each time step.

Littman's FFQ algorithm can guarantee the convergence to a Nash equilibrium if all states and actions are visited infinitely often. The proof of convergence for the FFQ algorithm can be found in Reference 10. Similar to the minimax Q- and Nash Q-learning algorithms, the learning speed is low because of the execution of linear programming at each iteration in Algorithm 4.3.

Algorithm 4.3 Friend-or-foe Q-learning algorithm

Initialize $V_i(s) = 0$ and $Q_i(s, a_1, \dots, a_{n_1}, o_1, \dots, o_{n_2}) = 0$ where $(a_1, \dots, a_{n_1})$ denotes player i and its friends' actions and $(o_1, \dots, o_{n_2})$ denotes its opponents' actions.

for Each iteration **do**

    Player i takes an action a_i from current state s based on an exploration-exploitation strategy.

    At the subsequent state s', player i observes the received reward r_i, its friends' and opponents' actions taken at state s.

    Update $Q_i(s, a_1, \dots, a_{n_1}, o_1, \dots, o_{n_2})$:

$$Q_i(s, a_1, \dots, a_{n_1}, o_1, \dots, o_{n_2}) \leftarrow (1 - \alpha)Q_i(s, a_1, \dots, a_{n_1}, o_1, \dots, o_{n_2})$$
$$+ \alpha\left[r_i + \gamma V_i(s')\right]$$

    where α is the learning rate and γ is the discount factor.

    Update $V_i(s)$ using linear programming:

$$V_i(s) = \max_{\pi_1(s,\cdot), \dots, \pi_{n_1}(s,\cdot)} \min_{o_1, \dots, o_{n_2} \in O_1 \times \cdots \times O_{n_2}} \sum_{a_1, \dots, a_{n_1} \in A_1 \times \cdots \times A_{n_1}}$$
$$Q_i(s, a_1, \dots, a_{n_1}, o_1, \dots, o_{n_2})\pi_1(s, a_1) \cdots \pi_{n_1}(s, a_{n_1}) \quad (4.58)$$

end for

4.9 Infinite Gradient Ascent

It is known in the game theory literature that the strategies need not converge when they are computed by the gradient ascent in two-person iterated games. Singh et al. proved that the average payoffs of two players will always converge to the expected payoffs of some Nash equilibrium even if their strategies do not converge. Let a two-player two-action general-sum game be defined by the following matrices:

$$R = \begin{bmatrix} r_{11} & r_{12} \\ r_{21} & r_{22} \end{bmatrix} \quad \text{and} \quad C = \begin{bmatrix} c_{11} & c_{12} \\ c_{21} & c_{22} \end{bmatrix}$$

where R is the payoffs of the row player (player 1) and C is the payoffs of the column player (player 2). Assume that player 1 plays action i and player 2 plays action j. Thus, the payoffs that player 1 and player 2 get are R_{ij} and C_{ij}, respectively.

Player 1 and player 2 are said to follow a mixed strategy as they can choose their actions stochastically. Let us assume that $\alpha \in [0,1]$ is the probability of player 1 to choose action 1 and $(1 - \alpha)$ is the probability of player 1 to choose action 2. Let us also assume that $\beta \in [0,1]$ is the probability of player 2 to choose action 1 and $(1 - \beta)$ is the probability of player 2 to choose action 2. Thus, player 1's expected payoff to the strategy pair (α, β) is

$$V_r(\alpha, \beta) = r_{11}(\alpha\beta) + r_{22}((1 - \alpha)(1 - \beta)) + r_{12}(\alpha(1 - \beta)) + r_{21}((1 - \alpha)\beta) \quad (4.59)$$

and player 2's expected payoff to the strategy pair (α, β) is

$$V_c(\alpha, \beta) = c_{11}(\alpha\beta) + c_{22}((1 - \alpha)(1 - \beta)) + c_{12}(\alpha(1 - \beta)) + c_{21}((1 - \alpha)\beta) \quad (4.60)$$

The effect of changing a player's strategy is estimated by calculating the partial derivative of its expected payoff with respect to its mixed strategy as follows:

$$\frac{\partial V_r(\alpha, \beta)}{\partial \alpha} = \beta u - (r_{22} - r_{12}) \quad (4.61)$$

$$\frac{\partial V_c(\alpha, \beta)}{\partial \beta} = \alpha \acute{u} - (c_{22} - c_{21}) \quad (4.62)$$

where $u = (r_{11} + r_{22}) - (r_{21} + r_{12})$ and $\acute{u} = (c_{11} + c_{22}) - (c_{21} + c_{12})$.

In the gradient ascent algorithm, at each time step the current strategy of each player is adjusted in the direction of its current gradient with some step size η so as to maximize its expected payoff:

$$\alpha_{k+1} = \alpha_k + \eta \frac{\partial V_r(\alpha_k, \beta_k)}{\partial \alpha} \quad (4.63)$$

$$\beta_{k+1} = \beta_k + \eta \frac{\partial V_c(\alpha_k, \beta_k)}{\partial \beta} \quad (4.64)$$

The step size η is usually in the range $0 < \eta \ll 1$. It is clear that each player assumes that the opponent's strategy is known. Singh et al. proved that the agents, their average payoffs, or both of them will converge to a Nash equilibrium in case of the infinitesimal step size ($\lim_{\eta \to 0}$). However, the IGA (infinite gradient ascent) algorithm is not practical because it cannot be applied to a large number of real problems because of the following two reasons:

1. The opponent's strategy is assumed to be totally known by the player;

2. The IGA algorithm is designed for two-player two-action iterated general-sum games; for many-player many-action general-sum games, the extension will not be straightforward.

4.10 Policy Hill Climbing

Policy hill climbing (PHC) is a simple practical algorithm that can play mixed strategies. This algorithm was first proposed by Bowling and Veloso (2002). The PHC does not require much information as neither the player's recently executed actions nor its opponent's current strategy is required to be known. The PHC is a simple modification of the single-agent Q-learning algorithm. A hill climbing is performed by the PHC algorithm in the space of the mixed strategies. The PHC algorithm is composed of two parts. The reinforcement learning is the first part, as the Q-learning algorithm maintains the values of the particular actions in the states. The game-theoretic part is the second part in which the current strategy in each system's state is maintained.

The probability that selects the highest valued actions is increased by a small learning rate $\delta \in (0,1]$ so that the policy is improved. The algorithm is equivalent to Q-learning when $\delta = 1$, as the policy moves to the greedy policy with probability 1 while executing the highest valued action. The PHC algorithm is rational and converges to the optimal solution when a fixed (stationary) strategy is followed by the other players. However, the PHC algorithm may not converge to a stationary policy if the other players are learning although its average reward will converge to the reward of a Nash equilibrium. The PHC algorithm is illustrated in Algorithm 4.4.

4.11 WoLF-PHC Algorithm

Algorithm 4.4 Policy hill climbing algorithm

1: Initialize $Q_i(s, a_i) \leftarrow 0$ and $\pi_i(s, a_i) \leftarrow \dfrac{1}{|A_i|}$. Choose the learning rate α, δ and the discount factor γ.

2: **for** Each iteration **do**

3:   Select action a_c from current state s based on a mixed exploration-exploitation strategy

4:   Take action a_c and observe the reward r_i and the subsequent state s'

5:   Update $Q_i(s, a_c)$

$$Q_i(s, a_c) = Q_i(s, a_c) + \alpha \left[r_i + \gamma \max_{a_i'} Q(s', a_i') - Q(s, a_c) \right] \qquad (4.65)$$

where a_i' is player i's action at the next state s' and a_c is the action player i has taken at state s.

6:     Update $\pi_i(s, a_i)$

$$\pi_i(s, a_i) = \pi_i(s, a_i) + \Delta_{sa_i} \quad (\forall a_i \in A_i) \tag{4.66}$$

where

$$\Delta_{sa_i} = \begin{cases} -\delta_{sa_i} & \text{if } a_c \neq \arg\max_{a_i \in A_i} Q_i(s, a_i) \\ \sum_{a_j \neq a_i} \delta_{sa_j} & \text{otherwise} \end{cases} \tag{4.67}$$

$$\delta_{sa_i} = \min\left(\pi_i(s, a_i), \frac{\delta}{|A_i| - 1}\right) \tag{4.68}$$

7:     **end for**

The WoLF-PHC (win-or-learn-fast policy hill climbing) algorithm is an extension of the PHC algorithm [2]. This algorithm uses the mechanism of WoLF so that the PHC algorithm converges to a Nash equilibrium in self-play. The algorithm has two different learning rates: δ_w when the algorithm is winning, and δ_l when it is losing. The difference between the average strategy and the current strategy is used as a criterion to decide when the algorithm wins or loses. The learning rate δ_l is larger than the learning rate δ_w. As such, when an agent is losing, it learns faster than when it is winning. This causes the agent to adapt quickly to the changes in the strategies of the other agents when it is doing more poorly than expected, and learns cautiously when it is doing better than expected. This also gives the other agents the time to adapt to the agent's strategy changes. The WoLF-PHC algorithm exhibits the property of convergence as it makes the agent converge to one of its Nash equilibria. This algorithm is also a rational learning algorithm as it makes the agent converge to its optimal strategy when its opponent plays a stationary strategy. These properties permit the WoLF-PHC algorithm to be widely applied to a variety of stochastic games [2, 16–18]. The recursive Q-learning of a learning agent j is given as

$$Q_{t+1}^j(s, a) = (1 - \alpha)Q_t^j(s, a) + \alpha(r^j + \gamma \max_{a'} Q_t^j(s', a')) \tag{4.69}$$

Algorithm 4.5 describes the WoLF-PHC algorithm for a learning agent j, and the algorithm updates the strategy of agent j by the following equation:

$$\pi_{t+1}^j(s, a) = \pi_t^j(s, a) + \Delta_{sa} \tag{4.70}$$

where

$$\Delta_{sa} = \begin{cases} -\delta_{sa} & \text{if } a \neq \arg\max_{a'} Q_t^j(s, a') \\ \sum_{a' \neq a} \delta_{sa'} & \text{otherwise} \end{cases}$$

$$\delta_{sa} = \min\left(\pi_t^j(s, a), \frac{\delta}{|A_j| - 1}\right)$$

$$\delta = \begin{cases} \delta_w & \text{if } \sum_{a'} \pi_t(s, a') Q_{t+1}^j(s, a') > \sum_{a'} \bar{\pi}_{t+1}(s, a') Q_{t+1}^j(s, a') \\ \delta_l & \text{otherwise} \end{cases}$$

$$\bar{\pi}_{t+1}^j(s, a') = \bar{\pi}_t^j(s, a') + \frac{1}{C_{t+1}(s)}(\pi_t^j(s, a') - \bar{\pi}_t^j(s, a')) \quad \forall a' \in A_j$$

$$C_{t+1}(s) = C_t(s) + 1.$$

The WoLF-PHC algorithm for player i is provided in Algorithm 4.5.

Algorithm 4.5 WoLF-PHC learning algorithm

Initialize $Q_i(s, a_i) \leftarrow 0$, $\pi_i(s, a_i) \leftarrow \dfrac{1}{|A_i|}$ and $C(s) \leftarrow 0$.Choose the learning rate α, δ and the discount factor γ

for Each iteration **do**

    Select action a_c from current state s based on a mixed exploration-exploitation strategy

    Take action a_c and observe the reward r_i and the subsequent state s'

    Update $Q_i(s, a_c)$

$$Q_i(s, a_c) = Q_i(s, a_c) + \alpha\left[r_i + \gamma \max_{a_i'} Q(s', a_i') - Q(s, a_c)\right] \quad (4.71)$$

where a_i' is player i's action at the next state s' and a_c is the action player i has taken at state s.

Update the estimate of average strategy $\bar{\pi}_i$

$$C(s) = C(s) + 1 \quad (4.72)$$

$$\bar{\pi}_i(s, a_i) = \bar{\pi}_i(s, a_i) + \frac{1}{C(s)}\left(\pi_i(s, a_i) - \bar{\pi}_i(s, a_i)\right) \quad (\forall a_i \in A_i) \quad (4.73)$$

where $C(s)$ denotes how many times the state s has been visited.

Update $\pi_i(s, a_i)$

$$\pi_i(s, a_i) = \pi_i(s, a_i) + \Delta_{sa_i} \quad (\forall a_i \in A_i) \quad (4.74)$$

where

$$
\Delta_{sa_i} =
\begin{cases}
-\delta_{sa_i} & \text{if } a_c \neq \arg\max_{a_i \in A_i} Q_i(s, a_i) \\
\sum_{a_j \neq a_i} \delta_{sa_j} & \text{otherwise}
\end{cases}
\tag{4.75}
$$

$$
\delta_{sa_i} = \min\left(\pi_i(s, a_i), \frac{\delta}{|A_i| - 1}\right)
\tag{4.76}
$$

$$
\delta =
\begin{cases}
\delta_w & \text{if } \sum_{a_i \in A_i} \pi_i(s, a_i) Q_i(s, a_i) > \sum_{a_i \in A_i} \bar{\pi}_i(s, a_i) Q_i(s, a_i) \\
\delta_l & \text{otherwise}
\end{cases}
$$

end for

Different from the previously mentioned learning algorithms, the WoLF-PHC algorithm does not need to observe the other players' strategies and actions. Therefore, compared to the other three learning algorithms, the WoLF-PHC algorithm needs less information from the environment. Since the WoLF-PHC algorithm is based on the PHC method, neither linear programming nor quadratic programming is required in this algorithm. Since the WoLF-PHC algorithm is a practical algorithm, there was no proof of convergence provided in Reference 2. Instead, simulation results in Reference 2 illustrated the convergence of players' strategies by carefully choosing the learning rate according to different examples in matrix games and stochastic games.

4.12 Guarding a Territory Problem in a Grid World

The game of guarding a territory was first introduced by Isaacs [3]. In the game, the invader tries to move to the territory as close as possible while the defender tries to intercept and keep the invader away from the territory as far as possible. The practical application of this game can be found in surveillance and security missions for autonomous mobile robots. There are a few published works in this field since the game was introduced [19, 20]. In these works, the defender tries to use a fuzzy controller to locate the invader's position [19] or applies a fuzzy reasoning strategy to capture the invader [20]. However, all these works assume that the defender knows its optimal policy and the invader's policy. There is no learning technique applied to the players in their works. In our work, we assume that the defender or the invader has no prior knowledge of his optimal policy and the opponent's policy. We apply learning algorithms to the players and let the defender or the invader obtain its own optimal behavior after learning.

The problem of guarding a territory in Reference 3 is a differential game problem where the dynamic equations of the players are typically differential equations. In our work, we will investigate how the players learn to behave with no knowledge of the optimal strategies. Therefore, the above problem becomes a multiagent learning problem in a multiagent system. In the literature, there are a number of papers on multiagent systems [21, 22]. Among the multiagent learning applications, the predator–prey or the pursuit problem in a grid world has been well studied [22, 23]. To better understand the learning process of the two players in the game, we create a grid game of guarding a territory, which has not been studied so far.

Most multiagent learning algorithms are based on MARL methods [22]. According to the definition of the game in Reference 3, the grid game we established is a two-player zero-sum stochastic game. The minimax-Q algorithm [1] is well suited to solve our problem. However, if the player does not always take the action that is most damaging to the opponent, the opponent might have better performance using a rational learning algorithm than the minimax-Q [21]. The rational learning algorithm we use here is the WoLF-PHC learning algorithm. In this section, we run simulations and compare the learning performance of the minimax-Q and WoLF-PHC algorithms.

The problem of guarding a territory in this section is the grid version of the guarding a territory game in Reference 3. The game is defined as follows:

- We take a 6 × 6 grid as the playing field, as shown in Fig. 4-13. The invader starts from the upper-left corner and tries to reach the territory before the capture. The territory is represented by a cell named T in Fig. 4-13. The defender starts from the bottom and tries to intercept the invader. The initial positions of the players are not fixed and can be chosen randomly.

- Both players can move up, down, left, or right. At each time step, both players take their actions simultaneously and move to their adjacent cells. If the chosen action will take the player off the playing field, the player will stay at the current position.

- The nine gray cells centered around the defender, shown in Fig. 4-13b, is the region where the invader will be captured. A successful invasion by the invader is defined as the situation where the invader reaches the territory before the capture or the capture happens at the territory. The game ends when the defender captures the invader or a successful invasion by the

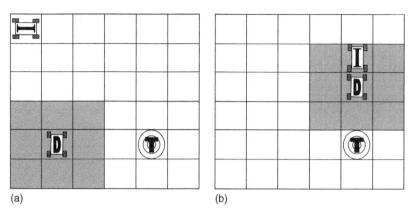

(a)   (b)

Fig. 4-13. Guarding a territory in a grid world. (a) Initial positions of the players when the game starts. (b) Terminal positions of the players when the game ends. Reproduced from [5], © X. Lu.

invader happens. Then the game restarts with random initial positions of the players.

- The goal of the invader is to reach the territory without interception, or move to the territory as close as possible if the capture must happen. On the contrary, the aim of the defender is to intercept the invader at a location as far away as possible from the territory.

The terminal time is defined as the time when the invader reaches the territory or is intercepted by the defender. We define the payoff as the distance between the invader and the territory at the terminal time.

$$\text{Payoff} = |x_I(t_f) - x_T| + |y_I(t_f) - y_T| \qquad (4.77)$$

where $(x_I(t_f), y_I(t_f))$ is the invader's position at the terminal time t_f and (x_T, y_T) is the territory's position. Based on the definition of the game, the invader tries to minimize the payoff while the defender tries to maximize the payoff.

4.12.1   Simulation and Results

We use the minimax-Q and WoLF-PHC algorithms introduced in Sections 4.3 and 4.11 to simulate the grid game of guarding a territory. We first present a simple 2×2 grid game to explore the issues of mixed strategy, rationality, and convergence. Next, we enlarge the playing field to a 6×6 grid and examine the performance of the learning algorithms based on this large grid.

We set up two simulations for each grid game. In the first simulation, the players in the game use the same learning algorithm to play against each

Table 4.7   Comparison of multiagent reinforcement learning algorithms.

Algorithms	Applicability	Rationality	Convergence
Minimax-Q	Zero-sum SGs	No	Yes
Nash-Q learning	Specific general-sum SGs	No	Yes
Friend-or-foe Q learning	Specific general-sum SGs	No	Yes
WoLF-PHC	General-sum SGs	Yes	No

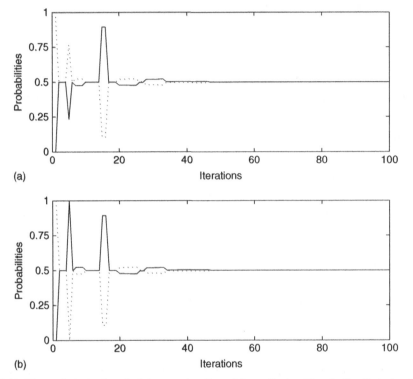

Fig. 4-14. Players' strategies at state s_1 using the minimax-Q algorithm in the first simulation for the 2 × 2 grid game. (a) Defender's strategy $\pi_D(s_1, a_{left})$ (solid line) and $\pi_D(s_1, a_{up})$ (dash line). (b) Invader's strategy $\pi_I(s_1, o_{down})$ (solid line) and $\pi_I(s_1, o_{right})$ (dash line). Reproduced from [5], © X. Lu.

other. We examine whether the algorithm satisfies the convergence property Fig. 4-14. In the second simulation, we will freeze one player's strategy and let the other player learn the optimal strategy against its opponent. We use the minimax-Q and WoLF-PHC algorithms to train the learner individually and compare the performance of the minimax-Q-trained player and the WoLF-PHC-trained player. According to the rationality property shown in Table 4.7, we expect the WoLF-PHC-trained defender has better performance than the minimax-Q-trained defender in the second simulation.

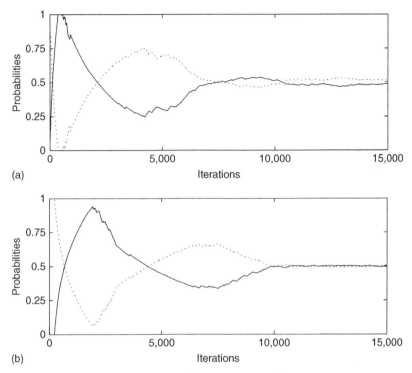

Fig. 4-15. Players' strategies at state s_1 using the WoLF-PHC algorithm in the first simulation for the 2×2 grid game. (a) Defender's strategy $\pi_D(s_1, a_{left})$ (solid line) and $\pi_D(s_1, a_{up})$ (dash line). (b) Invader's strategy $\pi_I(s_1, o_{down})$ (solid line) and $\pi_I(s_1, o_{right})$ (dash line). Reproduced from [5], © X. Lu.

We now apply the WoLF-PHC algorithm to the 2×2 grid game. According to the parameter settings in Reference 2, we set the learning rate α as $1/(10 + t/10000)$, δ_w as $1/(10 + t/2)$, and δ_l as $3/(10 + t/2)$, where t is the number of the current iteration. The number of iterations denotes the number of times the step 2 is repeated in Algorithm 4.5. The result in Fig. 4-15 shows that the players' strategies converge close to the Nash equilibrium after 15,000 iterations.

In the second simulation, the invader plays a stationary strategy against the defender at state s_1, as shown in Fig. 4-2a. The invader's fixed strategy is to move right with probability 0.8 and move down with probability 0.2. Then the optimal strategy for the defender against this invader is to move up all the time. We apply both algorithms to the game and examine the learning performance for the defender. Figure 4-16a shows that, using the minimax-Q algorithm, the defender's strategy fails to converge to its optimal strategy, whereas

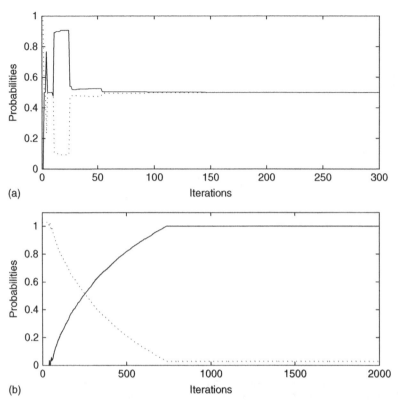

(a)

(b)

Fig. 4-16. Defender's strategy at state s_1 in the second simulation for the 2×2 grid game. (a) Minimax-Q-learned strategy of the defender at state s_1 against the invader using a fixed strategy. Solid line: probability of defender moving up; Dashed line: probability of defender moving left. (b) WoLF-PHC learned strategy of the defender at state s_1 against the invader using a fixed strategy. Solid line: probability of defender moving up; Dashed line: probability of defender moving left. Reproduced from [5], © X. Lu.

Fig. 4-16b shows that the WoLF-PHC algorithm guarantees the convergence to the defender's optimal strategy against the invader.

In the 2×2 grid game, the first simulation verified the convergence property of the minimax-Q and WoLF-PHC algorithms. According to Table 4.7, there is no proof of convergence for the WoLF-PHC algorithm. But simulation result in Fig. 4-15 shows that the players' strategies converged to the Nash equilibrium when both players used the WoLF-PHC algorithm. Under the rationality criterion, the minimax-Q algorithm failed to converge to the defender's optimal strategy in Fig. 4-16a, while the WoLF-PHC algorithm showed the convergence to the defender's optimal strategy after learning.

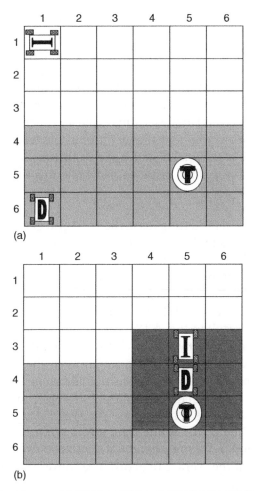

Fig. 4-17. A 6 × 6 grid game. (a) Initial positions of the players. (b) One of the terminal positions of the players. Reproduced from [5], © X. Lu.

We now change the 2 × 2 grid game to a 6 × 6 grid game. The territory to be guarded is represented by a cell located at (5, 5) in Fig. 4-17. The position of the territory will not be changed during the simulation. The initial positions of the invader and defender are shown in Fig. 4-17a. The number of actions for each player has been changed from 2 in the 2 × 2 grid game to 4 in the 6 × 6 grid game. Both players can move up, down, left, or right. The gray cells in Fig. 4-17a is the area where the defender can reach before the invader. Therefore, if both players play their equilibrium strategies, the invader can move to the territory as close as possible with the distance of two cells shown in Fig. 4-17b. Different from the previous 2 × 2 grid game where we showed the convergence of the players' strategies during the learning, in this game we want to show the average leaning performance of the players during the learning. We add a testing phase to evaluate the learned strategies after every 100 iterations. The number of iterations denotes the number of times step 2 is repeated in Algorithms 4.1 or 4.5. A testing phase includes 1000 runs of the game. In each run, the learned players start from their initial positions shown in Fig. 4-17a and end at the terminal time. For each run, we find the final distance between the invader and the territory at the terminal time. Then we calculate the average of the final distance over 1000 runs. The result of a testing phase, which is the average final distance over 1000 runs, is collected after every 100 iterations.

We use the same parameter settings as in the 2×2 grid game for the minimax-Q algorithm. In the first simulation, we test the convergence property by using the same learning algorithm for both players. Figure 4-18a shows the learning performance when both players used the minimax-Q algorithm. In Fig. 4-18a, the x-axis denotes the number of iterations and the y-axis denotes the result of the testing phase (the average of the final distance over 1000 runs) for every 100 iterations. From the result in Fig. 4-18a, the average final distance between the invader and the territory converges to 2 after 50,000 iterations. As shown in Fig. 4-17b, distance 2 is the final distance between the invader and the territory when both players play their Nash equilibrium strategies. Therefore, Fig. 4-18a indicates that both players' learned strategies converge close to their Nash equilibrium strategies. Then we use the WoLF-PHC algorithm to simulate again. We set the learning rate α as $1/(4 + t/50)$, δ_w as $1/(1 + t/5000)$, and δ_l as $4/(1 + t/5000)$. We run simulation for 200,000 iterations. The result in Fig. 4-18b shows that the average final distance converges close to the distance of 2 after the learning.

In the second simulation, we fix the invader's strategy to a random-walk strategy, which means that the invader can move up, down, left, or right with equal probability. Similar to the first simulation, the learning performance of the algorithms is tested on the basis of the result of a testing phase after every 100 iterations. In the testing phase, we play the game 1000 times and average the final distance between the invader and the territory at the terminal time for each run over 1000 runs.

We test the learning performance of both algorithms for the defender in the game and compare them. The results are shown in Fig. 4-19a and b. Using the WoLF-PHC algorithm, the defender can intercept the invader further away from the territory (distance of 6.6) than using the minimax-Q algorithm (distance of 5.9). Therefore, on the basis of the rationality criterion in

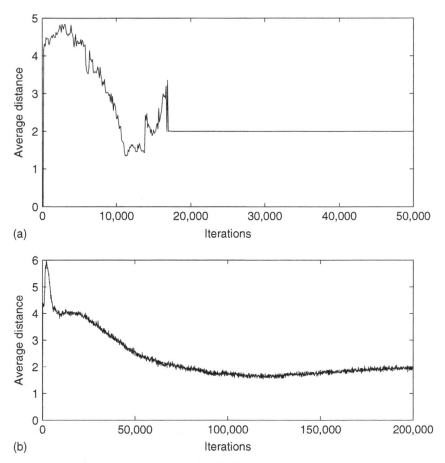

Fig. 4-18. Results in the first simulation for the 6 × 6 grid game. (a) Result of the minimax-Q learned strategy of the defender against the minimax-Q-learned strategy of the invader. (b) Result of the WoLF-PHC learned strategy of the defender against the WoLF-PHC-learned strategy of the invader. Reproduced from [5], © X. Lu.

Table 4.7, the WoLF-PHC-learned defender can achieve better performance than the minimax-Q-learned defender when playing against a random-walk invader.

4.13 Extension of L_{R-I} Lagging Anchor Algorithm to Stochastic Games

The proposed L_{R-I} lagging anchor algorithm is designed on the basis of matrix games. In this section, we extend the algorithm to the more general stochastic games. Inspired by the WoLF-PHC algorithm in Reference 2, we design a practical decentralized learning algorithm for stochastic games based on the

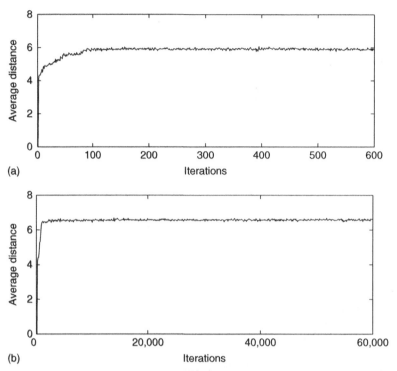

Fig. 4-19. Results in the second simulation for the 6 × 6 grid game. (a) Result of the minimax-Q-learned strategy of the defender against the invader using a fixed strategy. (b) Result of the WoLF-PHC-learned strategy of the defender against the invader using a fixed strategy. Reproduced from [5], © X. Lu.

L_{R-I} lagging anchor approach in (3.59). The practical algorithm is shown in Algorithm 4.6.

We now apply Algorithm 4.6 to a stochastic game to test its performance. The stochastic game we simulate is a general-sum grid game introduced by Hu and Wellman [8] and that we have already reviewed in section 4.4. Recall that. The game runs under a 3×3 grid field as shown in Fig. 4-20a. We have two players whose initial positions are located at the bottom left corner for player 1 and the bottom right corner for player 2. Both players try to reach the goal denoted as "G" in Fig. 4-20a. Each player has four possible moves which are moving up, down, left, or right unless the player is on the sides of the grid. In Hu and Wellman's game, the movement that will take the player to a wall is ignored. Since we use exactly the same game as Hu and Wellman, the possible actions of hitting a wall have been removed from the players' action sets. For example, if the player is at the bottom-left corner, its available moves are moving up or right. If both players move to the same cell at

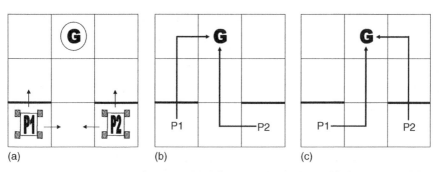

Fig. 4-20. Hu and Wellman's grid game. (a) Grid game. (b) Nash equilibrium path 1. (c) Nash equilibrium path 2. Reproduced from [24] © M. Awheda and Schwartz, H. M.

Algorithm 4.6 A practical L_{R-I} lagging anchor algorithm for player i

---

1: Initialize $Q_i(s, a_i) \leftarrow 0$ and $\pi_i(s, a_i) \leftarrow \dfrac{1}{|A_i|}$.Choose the learning rate α, η
   and the discount factor γ.
2: **for** Each iteration **do**
3:    Select action a_c at current state s based on mixed exploration-exploitation
      strategy
4:    Take action a_c and observe the reward r and the subsequent state s'
5:    Update $Q_i(s, a_c)$

$$Q_i(s, a_c) = Q_i(s, a_c) + \alpha\left[r_i + \gamma \max_{a'} Q_i(s', a') - Q_i(s, a_c)\right]$$

6:    Update the player's policy $\pi_i(s, \cdot)$

$$\pi_i(s, a_c) = \pi_i(s, a_c) + \eta Q_i(s, a_c)\left[1 - \pi_i(s, a_c)\right] + \eta\left[\bar{\pi}_i(s, a_c) - \pi_i(s, a_c)\right]$$

$$\bar{\pi}_i(s, a_c) = \bar{\pi}_i(s, a_c) + \eta\left[\pi_i(s, a_c) - \bar{\pi}_i(s, a_c)\right]$$

$$\pi_i(s, a_j) = \pi_i(s, a_j) - \eta Q_i(s, a_c)\pi_i(s, a_j) + \eta\left[\bar{\pi}_i(s, a_j) - \pi_i(s, a_j)\right]$$

$$\bar{\pi}_i(s, a_j) = \bar{\pi}_i(s, a_j) + \eta\left[\pi_i(s, a_j) - \bar{\pi}_i(s, a_j)\right]$$

      (for all $a_j \neq a_c$)

7: **end for**
   $\big(Q_i(s, a_i)$ is the action-value function, $\pi_i(s, a_i)$ is the probability of player i
   taking action a_i at state s and a_c is the current action taken by player i at
   state $s\big)$

---

the same time, they will bounce back to their original positions. The two thick lines in Fig. 4-20a represent two barriers such that the player can pass through the barrier with a probability of 0.5. For example, if player 1 tries to move up from the bottom-left corner, it will stay still or move to the upper cell with a probability of 0.5. The game ends when either of the players reaches the goal. To reach the goal in the minimum number of steps, the player needs to avoid the barrier and first move to the bottom center cell. Since both players cannot move to the bottom center cell simultaneously, the players need to cooperate such that one of the players has to take the risk and move up. The reward function for player i $(i = 1, 2)$ in this game is defined as

$$r_i = \begin{cases} 100 & \text{player } i \text{ reaches the goal} \\ -1 & \text{both players move to the same cell (except the goal)} \\ 0 & \text{otherwise} \end{cases} \qquad (4.78)$$

According to Reference 8, this grid game has two Nash equilibrium paths as shown in Fig. 4-20b and c. Starting from the initial state, the Nash equilibrium strategies of the players are player 1 moving up and player 2 moving left, or player 1 moving right and player 2 moving up.

We set the step size as $\eta = 0.001$, the learning rate as $\alpha = 0.001$, and the discount factor as $\gamma = 0.9$. The mixed exploration-exploitation strategy is chosen such that the player chooses a random action with probability 0.05 and the greedy action with probability 0.95. We run the simulation for 10,000 episodes. An episode is when the game starts with the players' initial positions and ends when either of the players reaches the goal. Figure 4-21 shows the result of two players' learning trajectories. We define p_1 as player 1's probability of moving up and q_1 as player 2's probability of moving up from their initial positions. The result in Fig. 4-21 shows that the two players' strategies at the initial state converge to one of the two Nash equilibrium strategies (player 1 moving right and player 2 moving up). Therefore, the proposed practical L_{R-I} lagging anchor algorithm may be applicable to general-sum stochastic games.

4.14 The Exponential Moving-Average Q-Learning (EMA Q-Learning) Algorithm

The exponential moving-average (EMA) approach is a model-free strategy estimation approach. It is a family of statistical approaches used to analyze time series data in finance and technical analysis. Typically, EMA gives the

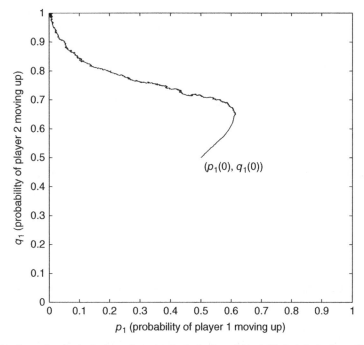

Fig. 4-21. Learning trajectories of players' strategies at the initial state in the grid game.
Reproduced from [5] © X. Lu.

recent observations more weight than the older ones [25]. The EMA estimation approach is used in Reference 26 by the hyper Q-learning algorithm to estimate the opponent's strategy. It is also used in Reference 25 by the IGA agent to estimate its opponent's strategy. The EMA estimator used to estimate the strategy of the agent's opponent(s) can be described by the following equation [25, 26]:

$$\pi_{t+1}^{-j}(s) = (1 - \eta)\pi_t^{-j}(s) + \eta\vec{u}(a^{-j}) \tag{4.79}$$

where $\pi^{-j}(s)$ is the opponent's strategy, η is a small constant step size $(0 < \eta \ll 1)$, and $\vec{u}(a^{-j})$ is a unit vector representation of the action a^{-j} chosen by the opponent $(-j)$ at the state s. The unit vector $\vec{u}(a^{-j})$ contains the same number of elements as π^{-j}. The elements in the unit vector $\vec{u}(a^{-j})$ are all equal to zero except for the element corresponding to the action a^{-j} which is equal to 1.

In this work, we are proposing a simple algorithm that uses the EMA approach. This algorithm is called the *EMA Q-learning algorithm*. The proposed algorithm uses the EMA mechanism as a basis to update the strategy of the agent itself. Furthermore, it uses two different variable learning rates

η_w and η_l when updating the agent's strategy instead of only one constant learning rate η used in References 25, 26. The values of these variable learning rates are inversely proportional to the number of iterations. The recursive Q-learning algorithm for a learning agent j is given by the following equation:

$$Q^j_{t+1}(s, a) = (1 - \theta)Q^j_t(s, a) + \theta(r^j + \zeta \max_{a'} Q^j_t(s', a')) \qquad (4.80)$$

Algorithm 4.7 The exponential moving-average (EMA) Q-learning algorithm for agent j

---

Initialize:
learning rates $\theta \in (0,1]$, η_l and $\eta_w \in (0,1]$
constant gain k
exploration rate ϵ
discount factor ζ
$Q^j(s, a) \leftarrow 0$ and $\pi^j(s) \leftarrow \frac{1}{|A_j|}$

Repeat
(a) From the state s select an action a according to the strategy $\pi^j_t(s, a)$ with some exploration.
(b) Observe the immediate reward r^j and the new state s'.
(c) Update $Q^j_{t+1}(s, a)$ using Eq. (4.80).
(d) Update the strategy $\pi^j_{t+1}(s, a)$ by using Eq. (4.81).

The EMA Q-learning algorithm updates the strategy of the agent j by Eq. (4.81), whereas Algorithm 4.7 lists the procedure of the EMA Q-learning algorithm for a learning agent j.

$$\pi^j_{t+1}(s) = (1 - k\eta)\pi^j_t(s) + k\eta\vec{u}(a) \qquad (4.81)$$

where

k is a constant gain.

$$\eta = \begin{cases} \eta_w & \text{if } a = \text{argmax}_{a'} Q^j_t(s, a') \\ \eta_l & \text{otherwise} \end{cases}$$

$$\vec{u}(a) = \begin{cases} \vec{u}(a^j) & \text{if } a = \text{argmax}_{a'} Q^j_t(s, a') \\ \vec{u}(a'^j) & \text{otherwise} \end{cases}$$

$\vec{u}(a^j)$ is a unit vector representation of the action a^j with zero elements except for the element corresponding to the action a^j which is equal to 1. This is to make the EMA Q-learning learn fast when the chosen action of agent j is equal to the greedy action obtained from the agent's Q-table. On the other hand, $\vec{u}(a'^j) = \frac{1}{|A_j|-1}[\vec{1} - \vec{u}(a^j)]$. This is to make the EMA Q-learning learn cautiously and increase the opportunity of exploring the other agent's actions when the chosen action of agent j and the greedy action obtained from the agent's Q-table are different.

4.15 Simulation and Results Comparing EMA Q-Learning to Other Methods

We have evaluated the EMA Q-learning, WoLF-PHC [2], GIGA-WoLF [27], weighted policy learning (WPL) [28], and policy gradient ascent with approximate policy prediction (PGA-APP) [29] algorithms on a variety of matrix and stochastic games. We only show the EMA Q-learning, the PGA-APP, and the WPL algorithms. The results of applying the WPL, PGA-APP, and EMA Q-learning algorithms to different matrix and stochastic games are presented in this section. A comparison among the three algorithms in terms of the convergence to Nash equilibrium is provided. We use the same learning and exploration rates for all algorithms when they are applied to the same game. In some cases, these rates are chosen to be close to those used in Reference 29. In other cases, the values of these rates are chosen on a trial-and-error basis to achieve the best performance of all algorithms.

4.15.1 Matrix Games

We revisit matrix games to illustrate the improved performance of the EMA Q-learning algorithm. The EMA Q-learning, PGA-APP, and WPL algorithms are applied to the matrix games. They are also applied to the three-player matching pennies game. Figure 4-22 shows the probability distributions of the second actions for both players in the dilemma game. The EMA Q-learning, PGA-APP, and WPL algorithms are shown. In this game, the parameters of the EMA Q-learning algorithm are set as follows: $\eta_w = \frac{1}{10+i/5}$, $\eta_l = 0.01\eta_w$, $k = 1$, $\zeta = 0$, and $\theta = 0.05$ with an exploration rate $\varepsilon = 0.05$. The parameter γ is set as $\gamma = 0.5$ in the PGA-APP algorithm and the learning rate η is decayed with a slower rate in the WPL algorithm and is set as $\eta = \frac{1}{10+i/350}$. Figure 4-23 shows the probability distributions of the first actions for the three players in the three-player matching pennies game while learning with the EMA Q-learning, PGA-APP, and WPL algorithms. In this game, the

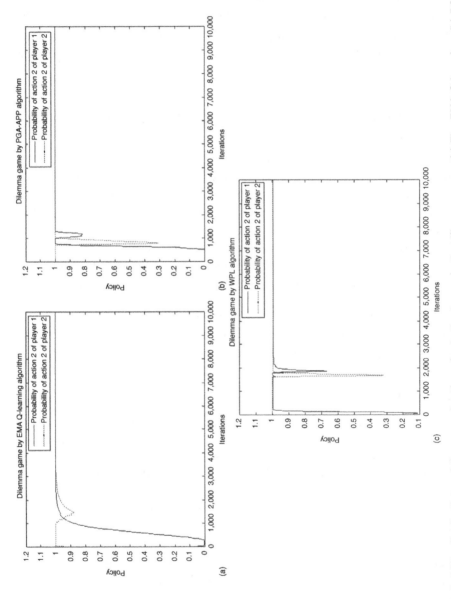

Fig. 4-22 Probability distributions of the second actions for both players in the dilemma game. (a) The EMA Q-learning, (b) PGA-APP, and (c) WPL algorithms are shown. Reproduced from [24] © M. Awheda and Schwartz, H. M.

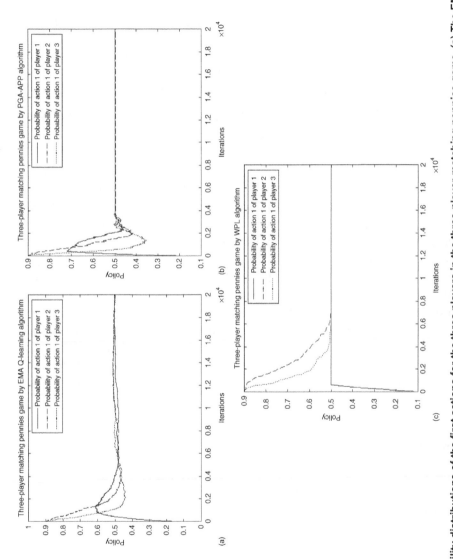

Fig. 4-23 Probability distributions of the first actions for the three players in the three-player matching pennies game. (a) The EMA Q-learning, (b) PGA-APP, and (c) WPL algorithms are shown. Reproduced from [24] © M. Awheda and Schwartz, H. M.

parameters of the EMA Q-learning algorithm are set as follows: $\eta_w = \frac{5}{5000+4i}$, $\eta_l = 2\eta_w$, $k = 1$, $\zeta = 0$, and $\theta = 0.8$ with an exploration rate $\varepsilon = 0.05$. The value of γ is set to $\gamma = 3$ in the PGA-APP algorithm and the learning rate η is decayed slowly in the WPL algorithm and is set as η as $\eta = \frac{5}{5000+i/500}$. Figure 4-24 shows the probability distributions of player 1's actions in the Shapley's game while learning with the EMA Q-learning, PGA-APP, and WPL algorithms. In this game, the parameters of the EMA Q-learning algorithm are set as follows: $\eta_w = \frac{1}{50+i}$, $\eta_l = 2\eta_w$, $k = 1$, $\zeta = 0$, and $\theta = 0.8$ with an exploration rate $\varepsilon = 0.05$. The learning rate η and the parameter γ in the PGA-APP algorithm are set as follows: $\eta = \frac{1}{50+i/50}$ and $\gamma = 3$. On the other hand, in the WPL algorithm, the learning rate η is decayed slowly and is set as $\eta = \frac{1}{50+i/200}$. Figure 4-25 shows the probability distributions of the first actions for both players in the biased game while learning with the EMA Q-learning, PGA-APP, and WPL algorithms. In this game, the parameters of the EMA Q-learning algorithm are set as follows: $\eta_w = \frac{1}{10+i/5}$, $\eta_l = 0.01\eta_w$, $k = 1$, $\zeta = 0.95$, and $\theta = 0.8$ with an exploration rate $\varepsilon = 0.05$. In the PGA-APP algorithm, the values of ζ and γ are set as follows: $\zeta = 0$ and $\gamma = 3$. In the WPL algorithm, the parameter ζ is set as $\zeta = 0$ and the learning rate η is decayed with a slower rate and is set as $\eta = \frac{1}{10+i/350}$. As shown in Figs. 4-22–4-24, the players' strategies successfully converge to Nash equilibria in all games when learning with the EMA Q-learning, PGA-APP, and WPL algorithms. It is important to mention here that the WPL algorithm successfully converges to Nash equilibrium in the three-player matching pennies game although it was presented to diverge in this game in Reference 29. On the other hand, Fig. 4-26 shows that both PGA-APP and WPL algorithms fail to converge to a Nash equilibrium in the biased game; only the EMA Q-learning algorithm succeeds to converge to a Nash equilibrium in the biased game.

4.15.2 Stochastic Games

It is important to mention here that we are only concerned about the first movement of both players from the initial state. Therefore, the figures that will be shown in this section will represent the probabilities of players' actions at the initial state.

4.15.2.1 Grid Game 1

The EMA Q-learning, PGA-APP, and WPL algorithms are used to learn grid game 1 depicted again in Fig. 4-27a. Grid game 1 has 10 different Nash equilibria [8]. One of these Nash equilibria is shown in Fig. 4-28a. Figure 4-28a

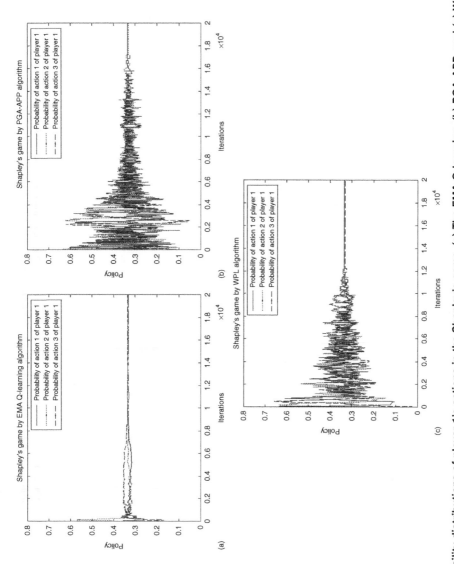

Fig. 4-24 Probability distributions of player 1's actions in the Shapley's game. (a) The EMA Q-learning, (b) PGA-APP, and (c) WPL algorithms are shown. Reproduced from [24] © M. Awheda and Schwartz, H. M.

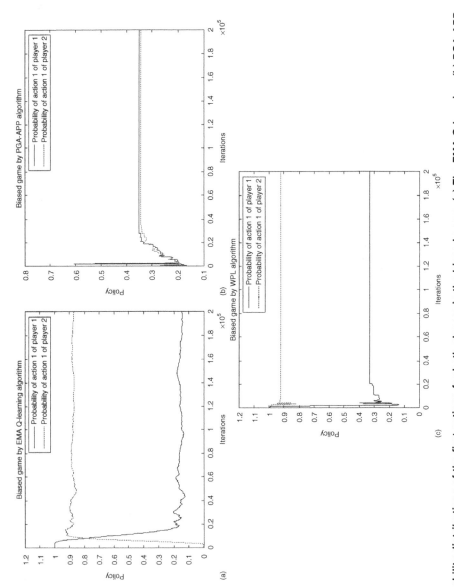

Fig. 4-25 Probability distributions of the first actions for both players in the biased game. (a) The EMA Q-learning, (b) PGA-APP, and (c) WPL algorithms are shown. Reproduced from [24] © M. Awheda and Schwartz, H. M.

136

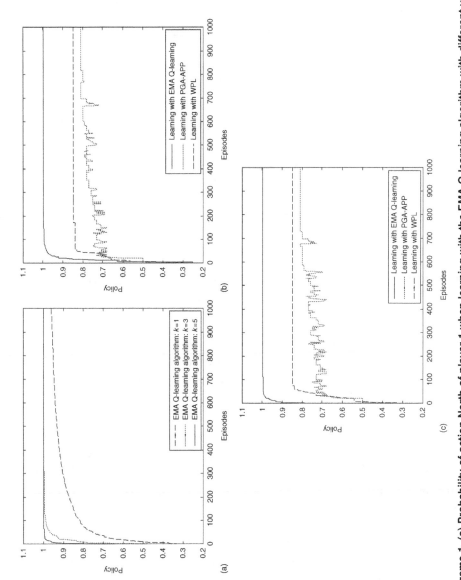

Fig. 4-26 Grid game 1. (a) Probability of action North of player 1 when learning with the EMA Q-learning algorithm with different values of the constant gain *k*. Plots (b) and (c) illustrate the probability of action North of player 1 and player 2, respectively, when learning with the EMA Q-learning, PGA-APP, and WPL algorithms. Reproduced from [24] © M. Awheda and Schwartz, H. M.

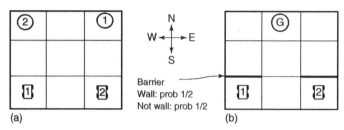

Fig. 4-27. Two stochastic games [8]. (a) Grid game 1. (b) Grid game 2. Reproduced from [24] ©
M. Awheda and Schwartz, H. M.

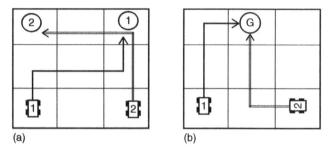

Fig. 4-28. (a) Nash equilibrium of grid game 1. (b) Nash equilibrium of grid game 2 [8] with
permission from MIT press. Reproduced from [24] © M. Awheda and Schwartz, H. M.

shows that the action North is the optimal action for both players when they are at the initial state. The learning and exploration rates used by all algorithms are the same. The parameters of the EMA Q-learning algorithm are set as follows: $\eta_w = \frac{1}{10+i}$, $\eta_l = 0.001\eta_w$, $k = 5$, $\zeta = 0$, and $\theta = 0.8$ with an exploration rate $\varepsilon = \frac{1}{1+0.001i}$, where i is the current number of episodes. The values of the parameters of the PGA-APP algorithm are the same as those of the EMA Q-learning algorithm except that $\gamma = 3$ and η has a very slow decaying rate, $\eta = \frac{1}{10+i/5000}$. The WPL algorithm also has the same parameters as the EMA Q-learning algorithm except that the learning rate η has a very slow decaying rate of $\eta = \frac{1}{10+i/5000}$.

Figure 4-26a shows the probability of selecting action North by player 1 at the initial state when learning with the EMA Q-learning algorithm with different values of the constant gain k. Player 2 has similar probability distributions when learning with the EMA Q-learning algorithm with different values of the constant gain k. Figure 4-26a shows that player 1's speed of convergence to the optimal action (North) increases as the value of the constant gain k increases. Figure 4-26a shows that the probability of selecting the optimal action North by player 1 requires almost 80 episodes to converge to 1 when $k=5$ and 320 episodes when $k=3$. However, when $k = 1$, many more episodes are still required for the probability of selecting the action North to converge to 1. Figure 4-26b and c shows the probabilities of taking action North at the initial state by both players when learning with the EMA Q-learning, PGA-APP, and WPL algorithms. This figure shows that the probabilities of taking action North by both players converge to the Nash equilibria (converge to 1) when learning with the EMA Q-learning algorithm. However, the PGA-APP and WPL algorithms fail to make the players' strategies converge to the Nash equilibria. Figure 4-26 shows that the EMA Q-learning algorithm outperforms the PGA-APP and WPL algorithms in terms of the convergence to Nash equilibria. It also shows that the EMA Q-learning algorithm can converge to Nash equilibria with a small number of episodes by adjusting the value of the constant gain k. This will give the EMA Q-learning algorithm an empirical advantage over the PGA-APP and WPL algorithms.

4.15.2.2 Grid Game 2

The EMA Q-learning, PGA-APP, and WPL algorithms are also used to learn grid game 2 as depicted in Fig. 4-27b. Grid game 2 has two Nash equilibria [8]. Figure 4-28b shows one of these Nash equilibria. It is apparent from this particular Nash equilibrium that the action North is the optimal action for player 1 at the initial state, whereas the action West is the optimal action for player 2. Thus, for the algorithms to converge to this particular Nash equilibrium, the probability of selecting the action North by player 1 should converge to 1. The probability of selecting the action West by player 2, on the other hand, should also converge to 1. The learning and exploration rates used by all algorithms are the same. The parameters of the EMA Q-learning algorithm are set as follows: $\eta_w = \frac{1}{10+i}$, $\eta_l = 0.001\eta_w$, $k = 10$, $\zeta = 0.1$, and $\theta = \frac{1}{1+0.001i}$ with an exploration rate $\varepsilon = \frac{1}{1+0.001i}$, where i is the current number of episodes. The values of the parameters of the PGA-APP algorithm are the same as those of the EMA Q-learning algorithm except that $\gamma = 3$, $\zeta = 0$, and η has a very slow decaying rate of $\eta = \frac{1}{10+i/5000}$. The WPL algorithm also has the same parameters as the EMA Q-learning algorithm except that $\zeta = 0$ and η has a very slow decaying rate of $\eta = \frac{1}{10+i/5000}$.

Figure 4-29a shows the probability of selecting action North by player 1 when learning with the EMA Q-learning, PGA-APP, and WPL algorithms. Figure 4-29a illustrates that the probability of selecting the action North by player 1 successfully converges to 1 (Nash equilibrium) when player 1 learns with the EMA Q-learning algorithm. However, the PGA-APP and WPL

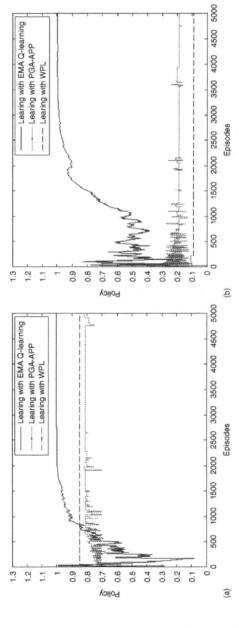

Fig. 4-29 Grid game 2. (a) Probability of selecting action North by player 1 when learning with the EMA Q-learning, PGA-APP, and WPL algorithms. (b) Probability of selecting action West by player 2 when learning with the EMA Q-learning, PGA-APP, and WPL algorithms. Reproduced from [24] © M. Awheda and Schwartz, H. M.

algorithms fail to make player 1 choose action North with a probability of 1. Figure 4-29b shows the probability of selecting action West by player 2 when learning with the EMA Q-learning, PGA-APP, and WPL algorithms. As can be seen from Fig. 4-29b, the probability of selecting action West by player 2 successfully converges to 1 (Nash equilibrium) when player 2 learns with the EMA Q-learning algorithm. The PGA-APP and WPL algorithms, on the other hand, fail to make player 2 choose action West with a probability of 1. Figure 4-29 shows that the EMA Q-learning algorithm outperforms the PGA-APP and WPL algorithms in terms of the convergence to Nash equilibria. This will give the EMA Q-learning algorithm an empirical advantage over the PGA-APP and WPL algorithms.

References

[1] M. L. Littman, "Markov games as a framework for multi-agent reinforcement learning," in *11th International Conference on Machine Learning*, (New Brunswick, United States), July 1994, pp. 157–163, 1994.

[2] M. Bowling and M. Veloso, "Multiagent learning using a variable learning rate," *Artificial Intelligence*, vol. 136, no. 2, pp. 215–250, 2002.

[3] R. Isaacs, *Differential Games: A Mathematical Theory with Applications to Warfare and Pursuit, Control and Optimization.* New York, New York: John Wiley and Sons, Inc., 1965.

[4] M. Bowling, Multiagent Learning in the Presence of Agents with Limitations. PhD thesis, School of Computer Science, Carnegie Mellon University, Pittsburgh, PA, May 2003.

[5] X. Lu, "On Multi-Agent Reinforcement Learning in Games," Ph.D. Thesis Carleton University, Ottawa, ON, Canada, 2012.

[6] M. L. Littman and C. Szepesvári, "A generalized reinforcement-learning model: Convergence and applications," in Proceedings of the 13th International Conference on Machine Learning, (Bari, Italy), July 1996, pp. 310–318, 1996.

[7] J. Hu and M. P. Wellman, "Multiagent reinforcement learning: theoretical framework and an algorithm," in Proceedings of the Fifteenth International Conference on Machine Learning (ICML 1998), Madison, Wisconsin, USA, July 24-27, 1998, pp. 242–250, 1998.

[8] J. Hu and M. P. Wellman, "Nash q-learning for general-sum stochastic games," *Journal of Machine Learning Research*, vol. 4, pp. 1039–1069, 2003.

[9] S. Abdallah, "Equilibrium in a stochastic n-person game," *Journal of Science in Hiroshima University, Series A-I*, 28: 89–93, 1964.

[10] M. L. Littman, "Friend-or-foe q-learning in general-sum games," in Proceedings of the 18th International Conference on Machine Learning, (Williams College, MA), pp. 322–328, 2001.

[11] C. Claus and C. Boutilier, "The dynamics of reinforcement learning in cooperative multiagent systems," in Proceedings of National Conference on Artificial Intelligence (AAAI-98), pp. 746–752, 1998.

[12] C. E. Lemke and J. J. T. Howson, "Equilibrium points of bimatrix games," *SIAM Journal on Applied Mathematics*, vol. 12, no. 2, pp. 413–423, 1964.

[13] D. D. Meredith, K. W. Wong, R. W. Woodhead, and R. H. Wortman, *Design and Planning of Engineering Systems*. Englewood Cliffs, New Jersey: Prentice-Hall, 1973.

[14] R. W. Cottle, J.-S. Pang, and R. E. Stone, "The linear complimentary problem," *Computer Science and Scientific Computing*, San Diego, California: Academic Press, Inc., 1992.

[15] P. De Beck-Courcelle, "Study of Multiple Multiagent Reinforcement Learning Algorithms in Grid Games", Master's thesis, Carleton University, Ottawa, ON, Canada, 2013.

[16] E. Yang and D. Gu, "A survey on multiagent reinforcement learning towards multi-robot systems," in *Proceedings of IEEE Symposium on Computational Intelligence and Games*, 2005.

[17] L. Buşoniu, R. Babuška, and B. D. Schutter, "Multiagent reinforcement learning: a survey," in *9th International Conference on Control, Automation, Robotics and Vision (ICARCV)*, pp. 1–6, 2006.

[18] X. Lu and H. M. Schwartz, "An investigation of guarding a territory problem in a grid world," in American Control Conference, pp. 3204–3210, 2010.

[19] K. H. Hsia and J. G. Hsieh, "A first approach to fuzzy differential game problem: guarding a territory," *Fuzzy Sets and Systems*, vol. 55, pp. 157–167, 1993.

[20] Y. S. Lee, K. H. Hsia, and J. G. Hsieh, "A strategy for a payoff-switching differential game based on fuzzy reasoning," *Fuzzy Sets and Systems*, vol. 130, no. 2, pp. 237–251, 2002.

[21] L. Buşoniu, R. Babuška, and B. D. Schutter, "A comprehensive survey of multiagent reinforcement learning," *IEEE Transactions on Systems, Man, and Cybernetics Part C*, vol. 38, no. 2, pp. 156–172, 2008.

[22] P. Stone and M. Veloso, "Multiagent systems: a survey from a machine learning perspective," *Autonomous Robots*, vol. 8, no. 3, pp. 345–383, 2000.

[23] J. W. Sheppard, "Colearning in differential games," *Machine Learning*, vol. 33, pp. 201–233, 1998.

[24] M. Awheda, and Schwartz, H.M., "Exponential Moving Average Q-Learning Algorithm", Proceedings of the IEEE Symposium Series on Computational Intelligence, Singapore, April 15–19, 2013.

[25] A. Burkov and B. Chaib-draa, "Effective learning in the presence of adaptive counterparts," *Journal of Algorithms*, vol. 64, no. 4, pp. 127–138, 2009.

[26] G. Tesauro, "Extending q-learning to general adaptive multi-agent systems," in *Advances in Neural Information Processing Systems 16* (S. Thrun, L. K. Saul and B. Schölkopf, eds.), (Cambridge, Massachusetts), pp. 215–250, MIT Press, 2004.

[27] M. Bowling, "Convergence and no-regret in multiagent learning," in *Advances in Neural Information Processing Systems 17* (L. K. Saul, Y. Weiss and L. Bottou, eds.), (Cambridge, Massachusetts), pp. 209–216, MIT Press, 2005.

[28] S. Abdallah and V. Lesser, "A multiagent reinforcement learning algorithm with non-linear dynamics," *Journal of Artificial Intelligence Research*, vol. 33, pp. 521–549, 2008.

[29] C. Zhang and V. Lesser, "Multi-agent learning with policy prediction," in Proceedings of the 24th National Conference on Artificial Intelligence (AAAI'10), Atlanta, GA, USA, pp. 746–752, 2010.

Chapter 5
Differential Games

5.1 Introduction

In the not too distant future, teams of robots will work together to accomplish a multitude of tasks. At the time of writing this book, we have seen the extensive use of aerial drones in surveillance, mapping, and other more unsavory tasks. We are also witnessing the beginning of truly autonomous vehicles for transportation. How long will it be before cars routinely drive themselves? We are currently on the verge of having multiple autonomous vehicles working together as some type of swarm. These groups of robots or autonomous vehicles will be a combination of aerial-, land-, and sea-based vehicles. These vehicles will have different configurations and capabilities. Unlike in the previous chapters, these vehicles will not be constrained to a grid, but, instead, they will be operating in a continuous and dynamically changing environment. The actions of these vehicles will be mathematically described by differential equations. The actions that the autonomous vehicles take will essentially and ultimately be control actions. These actions may be the setting of voltages on various actuators. We will refer to these types of systems as differential games (DGs).

The goal of these types of agents is to learn how to work together and how to adapt to changes in their own or other robots' capabilities. For example,

Multi-Agent Machine Learning: A Reinforcement Approach, First Edition. Howard M. Schwartz.
© 2014 John Wiley & Sons, Inc. Published 2014 by John Wiley & Sons, Inc.

if one or more of the other robots are disabled or destroyed, the remaining autonomous vehicles would have to adapt in real time to such a situation. Furthermore, the autonomous vehicles do not initially know the capabilities of the other robots, and each vehicle has to learn how to work with others.

In this chapter, we will use two well-known games to evaluate various methods of multiagent learning in these DGs. One of the games that we will examine is the "evader–pursuer" game and the other game is the "guarding a territory" game.

Future security applications will involve robots protecting critical infrastructure [1]. The robots will work together to prevent intruders from crossing a secured area. They will have to adapt to an unpredictable and continuously changing environment. Their goal is to learn what actions to take in order to get optimum performance in security tasks. We model this application as the "guarding a territory" game. The DG of guarding a territory was first introduced by Isaacs [2]. In this game, the invader tries to get as close as possible to the territory, while the defender tries to intercept and keep the invader as far away as possible from the territory. The Isaacs' guarding a territory game is a DG where the dynamic equations of the players are differential equations. In the pursuer–evader game, the pursuer tries to capture the evader, while the evader tries to escape from capture. The practical application of this game can be found in surveillance and security missions for autonomous mobile robots.

A player in a DG needs to learn what actions to take if there is no prior knowledge of its optimal strategy. Learning in DGs has attracted attention in References 3–6. In these articles, reinforcement learning algorithms are applied to the players in the pursuer–evader game. Early work on the guarding a territory game can be found in References 7, 8, but there is no investigation on how the players can learn their optimal strategies by playing the game. We assume the defender has no prior knowledge of its optimal strategy or the invader's strategy. We investigate how reinforcement learning algorithms can be applied to the DG of guarding a territory.

Traditional reinforcement learning algorithms such as Q-learning may lead to the curse of dimensionality problem because of the intractable, continuous state space and action space. To avoid this problem, one may use fuzzy systems to represent the continuous space [9]. Fuzzy reinforcement learning methods have been applied to the pursuer–evader DG in References 4–6. In Reference 5, we applied a fuzzy actor–critic learning (FACL) algorithm to the pursuit evasion game.

5.2　A Brief Tutorial on Fuzzy Systems

Fuzzy systems have been used in a wide variety of applications in engineering, science, business, medicine, psychology, and other fields [10]. In engineering for example, some potential application areas include [10] the following:

- Aircraft/spacecraft: flight control, engine control, avionic systems, failure diagnosis, navigation, and satellite attitude control;
- Robotics: position control and path planning;
- Autonomous vehicles: ground and underwater;
- Automated highway systems: automatic steering, braking, and throttle control for vehicles.

In this chapter, we make use of fuzzy inference systems (FISs) for the control of both the robots and the critic. The critic is an FIS that approximates the continuous Q-function over the continuous state and action spaces. Therefore, we provide a short tutorial of fuzzy systems in the following section.

5.2.1　Fuzzy Sets and Fuzzy Rules

Fuzzy sets use linguistic labels arranged by membership functions (MFs) to perform numerical computation [11]. Fuzzy set theory provides a way of dealing with information linguistically as an alternative to calculus.

The universe of discourse X is defined as a collection of elements x that have the same characteristics. A fuzzy set A in X can be denoted by Reference 11

$$A = \{(x, \mu_A(x))|x \in X\} \tag{5.1}$$

where $\mu_A(x)$ is the MF for the fuzzy set A. The MFs can have values between 0 and 1. The MF maps the elements of the universe of discourse to membership degrees between 0 and 1. If $\mu_A(x)$ has values of 0 or 1, the fuzzy set A is called a *crisp* or a *classical* set.

The interpretations of set operations, such as union and intersection, are complicated in fuzzy set theory because of the graded property of MFs. Zadeh [12] proposed the following definitions for union and intersection operations:

Union $\mu_{A \cup B}(x) = \max[\mu_A(x), \mu_B(x)]$

Intersection $\mu_{A \cap B}(x) = \min[\mu_A(x), \mu_B(x)]$

where A and B are fuzzy sets.

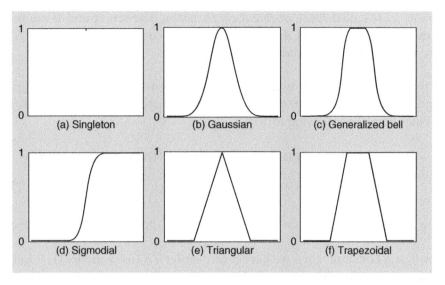

Fig. 5-1. Examples of membership functions. Reproduced from [13], © B, Al Faiya.

MFs are normally described using graphics. Figure 5-1 shows various types of MFs commonly used in fuzzy set theory. The Gaussian MF, for example, in Fig. 5-1b is given as

$$\mu_A(x) = \exp\left(-\left(\frac{x-m}{\sigma}\right)^2\right) \qquad (5.2)$$

where the Gaussian MF parameters are the mean m and the standard deviation σ.

The trapezoidal MF has four parameters, as shown in Fig. 5-1f. The trapezoidal MF is defined as

$$\mu(x) = \begin{cases} 0 & : \quad x < \alpha \\ \dfrac{x-\alpha}{\beta-\alpha} & : \alpha \le x < \beta \\ 1 & : \beta \le x \le \gamma \\ \dfrac{x-\gamma}{\lambda-\gamma} & : \gamma < x \le \lambda \\ 0 & : \quad x > \lambda \end{cases} \qquad (5.3)$$

Fuzzy IF-THEN rules can effectively model human expertise in an environment of uncertainty and imprecision [11]. Fuzzy IF-THEN rules are defined as

$$\mathfrak{R}_l : \textit{if } x \textit{ is } A \textit{ then } y \textit{ is } B \qquad (5.4)$$

where x, y are called *fuzzy* or *linguistic* variables. The sets A and B are fuzzy sets defined in X, Y. "x is A" is called the *antecedent* or *premise*, "y is B" is called the *consequence* or *conclusion*.

The fuzzy IF-THEN rules used in a Takagi–Sugeno (TS) fuzzy system give a mapping from the input fuzzy sets to a linear function in the output [14, 15]. The rules have the following form:

$$\mathfrak{R}_l : IF \ x_1 \ is \ A_1^l \ AND \ x_2 \ is \ A_2^l \ AND \ \ldots \ AND \ x_j \ is \ A_j^l \ THEN$$

$$f_l = K_0^l + \cdots + K_j^l x_j \tag{5.5}$$

where f_l is the output function of rule l and K_n^l is the consequent parameter.

When f_l is a constant, then we have a zero-order TS fuzzy model [11]. Let K^l be the constant for the output f_l. The number of rules is determined by the number of inputs and their corresponding MFs. Given two inputs and three MFs for each input, we need to construct nine ($3^2 = 9$) rules. The rules are given as follows:

$$\mathfrak{R}_1 : IF \ x_1 \ is \ A_1 \ AND \ x_2 \ is \ A_4 \ THEN \ f_1 = K^1$$
$$\mathfrak{R}_2 : IF \ x_1 \ is \ A_1 \ AND \ x_2 \ is \ A_5 \ THEN \ f_2 = K^2$$
$$\mathfrak{R}_3 : IF \ x_1 \ is \ A_1 \ AND \ x_2 \ is \ A_6 \ THEN \ f_3 = K^3$$
$$\mathfrak{R}_4 : IF \ x_1 \ is \ A_2 \ AND \ x_2 \ is \ A_4 \ THEN \ f_4 = K^4$$
$$\mathfrak{R}_5 : IF \ x_1 \ is \ A_2 \ AND \ x_2 \ is \ A_5 \ THEN \ f_5 = K^5$$
$$\mathfrak{R}_6 : IF \ x_1 \ is \ A_2 \ AND \ x_2 \ is \ A_6 \ THEN \ f_6 = K^6$$
$$\mathfrak{R}_7 : IF \ x_1 \ is \ A_3 \ AND \ x_2 \ is \ A_4 \ THEN \ f_7 = K^7$$
$$\mathfrak{R}_8 : IF \ x_1 \ is \ A_3 \ AND \ x_2 \ is \ A_5 \ THEN \ f_8 = K^8$$
$$\mathfrak{R}_9 : IF \ x_1 \ is \ A_3 \ AND \ x_2 \ is \ A_6 \ THEN \ f_9 = K^9.$$

Another format for constructing the fuzzy rules is the tabular format, as shown in Table 5.1.

5.2.2 Fuzzy Inference Engine

A fuzzy inference engine is used to combine fuzzy IF-THEN rules in the fuzzy rule base into a mapping from a fuzzy set A' in X to a fuzzy set B' in Y. One of the commonly used fuzzy inference engines is called the *product inference*

Table 5.1 Tabular format.

x_1 \ x_2	A_4	A_5	A_6
A_1	K^1	K^2	K^3
A_2	K^4	K^5	K^6
A_3	K^7	K^8	K^9

engine. In this section, the structure of the product inference engine is presented and explained.

We first provide the two operations on fuzzy sets: intersection and union. Assume we have two fuzzy sets A and B defined in the same universe of discourse U; the intersection of these two fuzzy sets is a fuzzy set whose MF is

$$\mu_{A \cap B}(x) = T(\mu_A(x), \mu_B(x)) = \mu_A(x) * \mu_B(x) \qquad (5.6)$$

where $*$ is defined as a t-norm operator. Two commonly used t-norm operators are the following:

$$\text{Minimum: } T_{\min}(a, b) = \min(a, b) \qquad (5.7)$$

$$\text{Algebraic product: } T_{ap}(a, b) = ab \qquad (5.8)$$

The union of two fuzzy sets A and B is a fuzzy set whose MF is given by

$$\mu_{A \cup B}(x) = S(\mu_A(x), \mu_B(x)) = \mu_A(x) \dotplus \mu_B(x) \qquad (5.9)$$

where $\dotplus$ is denoted as an s-norm operator. Two commonly used s-norm operators are as follows:

$$\text{Maximum: } S_{\max}(a, b) = \max(a, b) \qquad (5.10)$$

$$\text{Algebraic sum: } S_{ap}(a, b) = a + b - ab \qquad (5.11)$$

In the product inference engine, the algebraic product is used for all the t-norm operators and max is used for all the s-norm operators.

To interpret the IF-THEN operation, one can use Mamdani implication. In the Mamdani implication, a fuzzy IF-THEN rule can be considered as a binary fuzzy relation, given by

$$\mu_R(x, y) = \mu_{A \times B}(x, y) = \mu_{A \to B}(x, y) = \mu_A(x) * \mu_B(y) \qquad (5.12)$$

where $A \to B$ is used to interpret the fuzzy relation. If (5.8) is used as the t-norm operator $*$ in (5.12), then the latter is called the *Mamdani's product implication*. In fuzzy logic, the generalized modus ponens is defined as

$$
\begin{aligned}
&\text{premise 1(rule):} \quad \text{if } x \text{ is } A \text{ then } y \text{ is } B \\
&\text{premise 2(fact):} \quad x \text{ is } A' \\
&\text{conclusion:} \quad y \text{ is } B'.
\end{aligned}
$$

Based on the generalized modus ponens, the fuzzy set B' is inferred as

$$
\mu_{B'}(y) = \sup_{x \in X} T[\mu_{A'}(x), \mu_{A \to B}(x, y)] \tag{5.13}
$$

where $T[\cdot]$ denotes the t-norm operator and sup denotes the greatest element in the set.

In the individual rule-based inference, each fuzzy IF-THEN rule generates an individual output fuzzy set, and the whole output of the fuzzy inference engine is the combination of all the individual output fuzzy sets. In the product inference engine, we combine the individual output fuzzy sets by union.

Overall, the product inference engine includes the following three parts:

1. algebraic product for all the t-norm operators and max for all the s-norm operators;
2. Mamdani's product implication;
3. individual-rule-based inference with union combination.

Based on the above structure of the product inference engine, (5.13) becomes

$$
\mu_{B'}(y) = \max_{l=1}^{M} \mu_{B'_l}(y)
$$

$$
= \max_{l=1}^{M} [\sup_{x \in X} (\mu_{A'}(\mathbf{x}) \prod_{j=1}^{n} \mu_{A_j^l}(x_j) \mu_{B^l}(y))] \tag{5.14}
$$

We take an example here. Assume that we have two fuzzy IF-THEN rules with two antecedents for each rule such that

$$
\begin{aligned}
&\text{premise 1 (rule 1):} \quad \text{if } x_1 \text{ is } A_1^1 \text{ and } x_2 \text{ is } A_2^1 \text{ then } y \text{ is } B^1 \\
&\text{premise 2 (rule 2):} \quad \text{if } x_1 \text{ is } A_1^2 \text{ and } x_2 \text{ is } A_2^2 \text{ then } y \text{ is } B^2 \\
&\text{premise 3 (fact):} \quad x_1 \text{ is } A_1' \text{ and } x_2 \text{ is } A_2' \\
&\text{conclusion:} \quad y \text{ is } B'
\end{aligned} \tag{5.15}
$$

Then the output of the product inference engine for (5.15) becomes

$$\mu_{B'}(y) = \max_{l=1}^{2} \max_{x_1, x_2} [\mu_{A'_1}(x_1)\mu_{A'_2}(x_2) \prod_{j=1}^{2} \mu_{A'_i}(x_j)\mu_{B'}(y)] \tag{5.16}$$

5.2.3 Fuzzifier and Defuzzifier

Figure 5-2 shows the fuzzy system structure. The first block in the fuzzy system is the fuzzifier. The fuzzifier converts each input, which is a precise quantity, to degrees of membership in an MF [16]. The fuzzification block matches the input values with the conditions of the rules. The fuzzification determines how well the condition of each rule matches that particular input. There is a degree of membership for each linguistic term that applies to that input variable.

Defuzzification is the conversion of a fuzzy quantity into a precise quantity. The weighted average defuzzification method is frequently used in fuzzy applications since it is one of the most computationally efficient methods [17]. The weighted average defuzzification method is expressed as

$$f = \frac{\sum_{l=1}^{M}\left(\prod_{j=1}^{J}\mu^{A^l_j}(x_j)\right) f_l}{\sum_{l=1}^{M}\left(\prod_{j=1}^{J}\mu^{A^l_j}(x_j)\right)} \tag{5.17}$$

where J is number of inputs and M is number of rules.

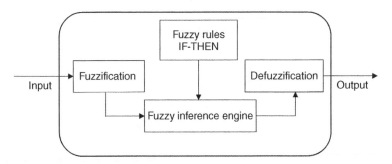

Fig. 5-2. Fuzzy system components. Reproduced from [13], © B, Al Faiya.

5.2.4   Fuzzy Systems and Examples

Fuzzy systems are also known as *FISs*, or *fuzzy controllers* when used as controllers. TS fuzzy systems and Mamdani fuzzy systems are commonly used in fuzzy applications. We want to investigate how well fuzzy systems can approximate a given system. The following theorem from Reference 15 is known as the "Universal Approximation Theorem." We will then give an example to show the capability of fuzzy systems.

Theorem 5.1   For any given real continuous function $g(\mathbf{x})$ on a compact set $U \subset R^n$ and arbitrary $\epsilon > 0$ with Gaussian MFs, there exists a fuzzy logic system $f(\mathbf{x})$ in the form of (5.17) such that

$$\sup_{\mathbf{x} \in U} |f(\mathbf{x}) - g(\mathbf{x})| < \epsilon \tag{5.18}$$

The proof is given in Reference 15 [pp. 124–126].

The next example introduced in Reference 18 shows the capability of fuzzy systems. It shows that fuzzy systems are good approximators of a given nonlinear system. The objective of the following example is to introduce the reader to FISs and to show how adding MFs can improve the approximation of nonlinear system by using an FIS.

Example 5.1   Consider a first-order nonlinear system. The dynamic equation of the system is given by Reference 15

$$\dot{x}(t) = \frac{1 - e^{-x(t)}}{1 + e^{-x(t)}} + u(t) = f(x) + u(t) \tag{5.19}$$

We define five fuzzy sets over the interval $[-3, 3]$: negative medium (NM), negative small (NS), zero (ZE), positive small (PS), and positive medium (PM). The MFs are given by

$$\mu_{NM}(x) = \exp(-(x + 1.5)^2)$$
$$\mu_{NS}(x) = \exp(-(x + 0.5)^2)$$
$$\mu_{ZE}(x) = \exp(-x^2)$$
$$\mu_{PS}(x) = \exp(-(x - 0.5)^2)$$
$$\mu_{PM}(x) = \exp(-(x - 1.5)^2) \tag{5.20}$$

Figure 5-3a shows the five MFs.

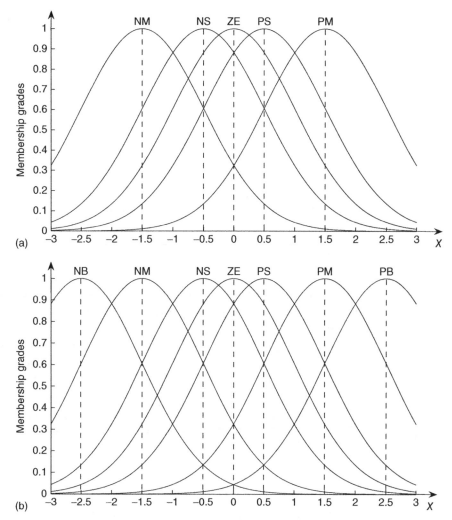

Fig. 5-3. Membership functions. (a) Membership functions of five fuzzy sets. (b) Membership functions of seven fuzzy sets. Reproduced from [13], © B, Al Faiya.

To estimate the dynamics of the system, the following linguistic descriptions (fuzzy IF-THEN rules) are given to the designer:

$$R^1: \quad \text{if } x \text{ is near } -1.5 \text{ then } f(x) \text{ is near } -0.6$$
$$R^2: \quad \text{if } x \text{ is near } -0.5 \text{ then } f(x) \text{ is near } -0.2$$
$$R^3: \quad \text{if } x \text{ is near } 0 \text{ then } f(x) \text{ is near } 0$$
$$R^4: \quad \text{if } x \text{ is near } 0.5 \text{ then } f(x) \text{ is near } 0.2$$
$$R^5: \quad \text{if } x \text{ is near } 1.5 \text{ then } f(x) \text{ is near } 0.6 \qquad (5.21)$$

We set the conclusions $y^1 = -0.6$, $y^2 = -0.2$, $y^3 = 0$, $y^4 = 0.2$, and $y^5 = 0.6$ as in (5.4). Since there is only one antecedent for every fuzzy IF-THEN rule, we rewrite (5.17) as

$$
\hat{f}(x) = \frac{\displaystyle\sum_{l=1}^{5} y^l [\mu_{A^l}(x)]}{\displaystyle\sum_{l=1}^{5} [\mu_{A^l}(x)]}
$$

$$
= \frac{-0.6\mu_{NM}(x) - 0.2\mu_{NS}(x) + 0.2\mu_{PS}(x) + 0.6\mu_{PM}(x)}{\mu_{NM}(x) + \mu_{NS}(x) + \mu_{ZE}(x) + \mu_{PS}(x) + \mu_{PM}(x)}
$$

$$
= \frac{-0.6e^{-(x+1.5)^2} - 0.2e^{-(x+0.5)^2} + 0.2e^{-(x-0.5)^2} + 0.6e^{-(x-1.5)^2}}{e^{-(x+1.5)^2} + e^{-(x+0.5)^2} + e^{-x^2} + e^{-(x-0.5)^2} + e^{-(x-1.5)^2}} \quad (5.22)
$$

To improve the performance of the fuzzy system, we need more specific MFs and linguistic descriptions in the fuzzy system. Therefore, we define seven fuzzy sets over the interval $[-3, 3]$: negative big (NB), negative medium (NM), negative small (NS), zero (ZE), positive small (PS), positive medium (PM), and positive big (PB). The MFs of the fuzzy sets NB and PB are defined as $\mu_{NB}(x) = \exp(-(x + 2.5)^2)$, $\mu_{PB}(x) = \exp(-(x - 2.5)^2)$. The MFs of fuzzy sets NM, NS, ZE, PS, and PM are the same as in (5.20). Figure 5-3b illustrates the MFs of the seven fuzzy sets. The fuzzy IF-THEN rules are given as follows:

$$R^1:\ \text{if } x \text{ is near } -2.5 \text{ then } f(x) \text{ is near } -0.85$$

$$R^2:\ \text{if } x \text{ is near } -1.5 \text{ then } f(x) \text{ is near } -0.64$$

$$R^3:\ \text{if } x \text{ is near } -0.5 \text{ then } f(x) \text{ is near } -0.24$$

$$R^4:\ \text{if } x \text{ is near } 0 \text{ then } f(x) \text{ is near } 0$$

$$R^5:\ \text{if } x \text{ is near } 0.5 \text{ then } f(x) \text{ is near } 0.24$$

$$R^6:\ \text{if } x \text{ is near } 1.5 \text{ then } f(x) \text{ is near } 0.64$$

$$R^7:\ \text{if } x \text{ is near } 2.5 \text{ then } f(x) \text{ is near } 0.85 \quad (5.23)$$

where $y^1 = -0.85$, $y^2 = -0.64$, $y^3 = -0.24$, $y^4 = 0$, $y^5 = 0.24$, $y^5 = 0.64$, and $y^5 = 0.85$, are the conclusions in (5.4).

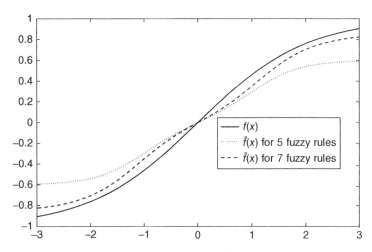

Fig. 5-4. Nonlinear function f(x) and the estimation f̂(x) with five rules and seven rules. Reproduced from [13], © B, Al Faiya.

Then the fuzzy system becomes

$$\hat{f}(x) = \frac{\sum_{l=1}^{7} y^l [\mu_{A^l}(x)]}{\sum_{l=1}^{7} [\mu_{A^l}(x)]}$$

$$= \frac{-0.85e^{-(x+2.5)^2} - 0.64e^{-(x+1.5)^2} - 0.24e^{-(x+0.5)^2} +}{e^{-(x+2.5)^2} + e^{-(x+1.5)^2} + e^{-(x+0.5)^2} + e^{-x^2} +}$$

$$\frac{0.24e^{-(x-0.5)^2} + 0.64e^{-(x-1.5)^2} + 0.85e^{-(x-2.5)^2}}{e^{-(x-0.5)^2} + e^{-(x-1.5)^2} + e^{-(x-2.5)^2}} \qquad (5.24)$$

Figure 5-4 shows the estimation $\hat{f}(x)$ (dashed line). Figure 5-5 shows the estimation error (solid line) $|f(x) - \hat{f}(x)|$ over the interval $[-3, 3]$.

5.3 Fuzzy Q-Learning

The value of the game is based on the assumption that both players play their Nash equilibrium strategies. In practical applications, one player may not know its own Nash equilibrium strategy or its opponent's strategy. Therefore, learning algorithms are needed to help the player learn its equilibrium strategy. Most of the learning algorithms applied to DGs, especially to the pursuer–evader game, are based on reinforcement learning algorithms [4–6].

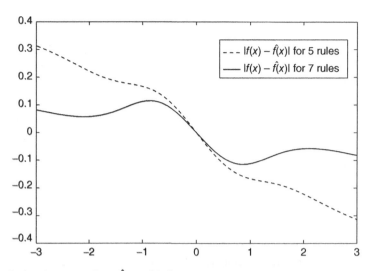

Fig. 5-5. Estimation error $|f(x) - \hat{f}(x)|$ with five rules and seven rules. Reproduced from [13], © B, Al Faiya.

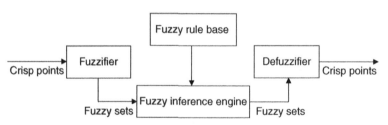

Fig. 5-6. Basic configuration of fuzzy systems. Reproduced from [13], with permission of Carleton University.

A typical reinforcement learning approach such as Q-learning needs to discretize the action space and the state space. However, when the continuous state space or action space is large, the discrete representation of the state or action is computationally intractable [19]. Wang [20] proved that an FIS is a universal approximator which can approximate any nonlinear function to any degree of precision. Therefore, one can use fuzzy systems to generate continuous actions of the players or represent the continuous state space.

The fuzzy system in this chapter, as shown in Fig. 5-6, is implemented by TS rules with constant consequents [21]. We structure the FIS for the use as part of a reinforcement learning system in the following way. It consists of M rules with n fuzzy variables as inputs and one constant number as the consequent.

Each rule l ($l = 1, \dots, M$) is of the form

$$R^l \; : \; \text{IF } x_1 \text{ is } F_1^l, \; \dots, \; \text{and } x_n \text{ is } F_n^l$$
$$\text{THEN } u = c^l \tag{5.25}$$

where $\bar{x} = (x_1, \dots, x_n)$ are the inputs passed to the fuzzy controller, F_i^l is the fuzzy set related to the corresponding fuzzy variable, u is the rule's output, and c^l is a constant that describes the center of a fuzzy set. If we use the product inference for fuzzy implication [20], t-norm, singleton fuzzifier, and center-average defuzzifier, the output of the system becomes

$$U(\bar{x}) = \frac{\displaystyle\sum_{l=1}^{M} \left(\left(\prod_{i=1}^{n} \mu^{F_i^l}(x_i) \right) \cdot c^l \right)}{\displaystyle\sum_{l=1}^{M} \left(\prod_{i=1}^{n} \mu^{F_i^l}(x_i) \right)} = \sum_{l=1}^{M} \Phi^l c^l \tag{5.26}$$

where $\mu^{F_i^l}$ is the membership degree of the fuzzy set F_i^l and

$$\Phi^l = \frac{\displaystyle\prod_{i=1}^{n} \mu^{F_i^l}(x_i)}{\displaystyle\sum_{l=1}^{M} \left(\prod_{i=1}^{n} \mu^{F_i^l}(x_i) \right)}. \tag{5.27}$$

Among fuzzy reinforcement learning algorithms, one may use a fuzzy Q-learning (FQL) algorithm to generate a global continuous action for the player based on a predefined discrete action set. We assume that the player has m possible actions from an action set $A = \{a_1, a_2, \dots, a_m\}$. To generate the player's global continuous action, we use the following form of fuzzy IF-THEN rules:

$$R_l \; : \; \text{IF } x_1 \text{ is } F_1^l, \; \dots, \; \text{and} \quad x_n \text{ is } F_n^l$$
$$\text{THEN } u = a^l \tag{5.28}$$

where a^l is the chosen action from the player's discrete action set A for rule l. The action a^l is chosen on the basis of an exploration-exploitation strategy [23]. In this chapter, we use the ε-greedy policy as the exploration-exploitation strategy. The ε-greedy policy is defined such that the player chooses a random action from the player's discrete action set A with a probability ε and a greedy action with a probability $1 - \varepsilon$. The greedy action is the action that gives the maximum value in an associated q-function. Then we have

$$a^l = \begin{cases} \text{random action from } A & \text{Prob}(\varepsilon) \\ \arg\max_{a \in A}(q(l, a)) & \text{Prob}(1 - \varepsilon) \end{cases} \tag{5.29}$$

where $q(l, a)$ is the associated q-function given the rule l and the player's action $a \in A$. Based on (5.26), the global continuous action at time t becomes

$$U_t(\bar{x}_t) = \sum_{l=1}^{M} \Phi_t^l a_t^l \tag{5.30}$$

where $\bar{x}_t = (x_1, x_2, \ldots, x_n)$ are the inputs, M is the number of fuzzy IF-THEN rules, and a_t^l is the chosen action in (5.29) for rule l at time t.

Similar to (5.30), we can generate the global Q-function by replacing c_l in (5.26) with $q_t(l, a_t^l)$ and get

$$Q_t(\bar{x}_t) = \sum_{l=1}^{M} \Phi_t^l q_t(l, a_t^l) \tag{5.31}$$

We can also define $Q_t^*(\bar{x}_t)$ as the global Q-function with the maximum q-value for each rule. Then (5.31) becomes

$$Q_t^*(\bar{x}_t) = \sum_{l=1}^{M} \Phi_t^l \max_{a \in A} q_t(l, a) \tag{5.32}$$

where $\max_{a \in A} q_t(l, a)$ denotes the maximum value of $q_t(l, a)$ for all $a \in A$ in rule l.

Given (5.31) and (5.32), we define the temporal difference (TD) error as

$$\tilde{\varepsilon}_{t+1} = r_{t+1} + \gamma Q_t^*(\bar{x}_{t+1}) - Q_t(\bar{x}_t) \tag{5.33}$$

where $\gamma \in [0, 1)$ is the discount factor and r_{t+1} is the received reward at time $t + 1$. Then the update law for the q-function is given as

$$q_{t+1}(l, a_t^l) = q_t(l, a_t^l) + \eta \tilde{\varepsilon}_{t+1} \Phi_t^l, \qquad (l = 1, \ldots, M) \tag{5.34}$$

where η is the learning rate.

The FQL learning algorithm is summarized in Algorithm 5.1.

Algorithm 5.1 FQL algorithm

1: Initialize $q(\cdot) = 0$ and $Q(\cdot) = 0$;
2: **for** Each time step **do**
3:     Choose an action for each rule based on (5.29) at time t;
4:     Compute the global continuous action $U_t(\bar{x}_t)$ in (5.30);
5:     Compute $Q_t(\bar{x}_t)$ in (5.31);
6:     Take the global action $U_t(\bar{x}_t)$ and run the game;
7:     Obtain the reward r_{t+1} and the new inputs $\bar{x}_{t+1}$ at time $t + 1$;
8:     Compute $Q_t^*(\bar{x}_{t+1})$ in (5.32);
9:     Compute the temporal difference error $\tilde{\varepsilon}_{t+1}$ in (5.33);
10:    Update $q_{t+1}(l, a_t^l)$ in (5.34) for $l = 1, \ldots, M$;
11: **end for**

5.4 Fuzzy Actor–Critic Learning

In FQL, one has to define the player's action set A based on the knowledge of the player's continuous action space. Suppose we do not know how large the action space is or the exact region the action space is in, the determination of the action set becomes difficult. Moreover, the number of elements in the action set will be prohibitively large when the action space is large. Correspondingly, the dimension of the q function in (5.34) will be intractably large. To avoid this, we present in this section an FACL method.

The actor–critic learning system contains two parts: one is to choose the optimal action for each state called the *actor*, and the other is to estimate the future system performance called the *critic*. Figure 5-7 shows the architecture of the actor–critic learning system. The actor is represented by an adaptive fuzzy controller which is implemented as an FIS. We also propose to implement the critic as an FIS. We have implemented the adaptive fuzzy critic in References 6, 24. We showed that the adaptive fuzzy critic in Reference 6 performed better than the neural network proposed in Reference 19. In the implementation proposed in this chapter, we only adapt the output parameters of the fuzzy system, whereas in Reference 6 the input and output parameters of the fuzzy system are adapted, which is a more complex adaptive algorithm. The reinforcement signal r_{t+1} is used to update the output parameters of the adaptive controller and the adaptive fuzzy critic, as shown in Fig. 5-7.

The actor is represented by an adaptive fuzzy controller which is implemented by TS rules with constant consequents. Then the output of the fuzzy controller

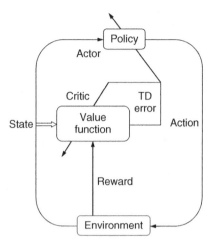

Fig. 5-7. Architecture of the actor–critic learning system. Reproduced from [13], with permission from MIT Press.

becomes

$$u_t = \sum_{l=1}^{M} \Phi^l w_t^l \tag{5.35}$$

where w^l is the output parameter of the actor.

In order to promote exploration of the action space, a random white noise $v(0, \sigma)$ is added to the generated control signal u. The output parameter of the actor w^l is adapted as

$$w_{t+1}^l = w_t^l + \beta \Delta(\frac{u_t' - u_t}{\sigma}) \frac{\partial u}{\partial w^l} \tag{5.36}$$

where $\beta \in (0, 1)$ is the learning rate for the actor.

In order to avoid large adaptation steps in the wrong direction [25], we use only the sign of the prediction error Δ and the exploration part $(u_t' - u_t)/\sigma$ in (5.36). Then, Eq. (5.36) becomes

$$w_{t+1}^l = w_t^l + \beta \text{sign} \left\{ \Delta \left(\frac{u_t' - u_t}{\sigma} \right) \right\} \frac{\partial u}{\partial w^l} \tag{5.37}$$

where

$$\frac{\partial u}{\partial w^l} = \frac{\prod_{i=1}^{n} \mu^{F_i^l}(x_i)}{\sum_{l=1}^{M} \left(\prod_{i=1}^{n} \mu^{F_i^l}(x_i) \right)} = \Phi_t^l \tag{5.38}$$

The task of the critic is to estimate the value function over a continuous state space. The value function is the expected sum of discounted rewards defined as

$$V_t = E\left\{\sum_{k=0}^{\infty} \gamma^k r_{t+k+1}\right\} \qquad (5.39)$$

where t is the current time step, r_{t+k+1} is the received immediate reward at the time step $t+k+1$, and $\gamma \in [0, 1)$ is a discount factor. Equation (5.39) can be also rewritten recursively as

$$V_t = r_{t+1} + \gamma V_{t+1} \qquad (5.40)$$

After each action selection from the actor, the critic evaluates the new state to determine whether things have gone better or worse than expected. For the critic in Fig. 5-7, we assume TS rules with constant consequents [25]. The output of the critic $\hat{V}$ is an approximation to V, given by

$$\hat{V}_t = \sum_{l=1}^{M} \Phi^l \varsigma_t^l \qquad (5.41)$$

where t denotes a discrete time step, ς_t^l is the output parameter of the critic defined as c^l in (5.25), and Φ^l is defined in (5.27).

Based on (5.40) and the above approximation $\hat{V}_t$, we can generate a prediction error Δ as

$$\Delta = r_{t+1} + \gamma \hat{V}_{t+1} - \hat{V}_t \qquad (5.42)$$

This prediction error is then used to train the critic. Supposing it has the parameter ς^l to be adapted, the adaptation law would then be

$$\varsigma_{t+1}^l = \varsigma_t^l + \alpha \Delta \frac{\partial \hat{V}}{\partial \varsigma^l} \qquad (5.43)$$

where $\alpha \in (0, 1)$ is the learning rate for the critic. We set $\beta < \alpha$, where β is given in (5.36), so that the actor will converge more slowly than the critic to prevent instability in the actor [24]. Also the partial derivative is easily calculated to be

$$\frac{\partial \hat{V}}{\partial \varsigma^l} = \frac{\prod_{i=1}^{n} \mu^{F_i^l}(x_i)}{\sum_{l=1}^{M} \left(\prod_{i=1}^{n} \mu^{F_i^l}(x_i)\right)} = \Phi^l \qquad (5.44)$$

The FACL algorithm is summarized in Algorithm 5.2.

Algorithm 5.2 FACL algorithm

1: Initialize $\hat{V} = 0$, $\zeta^l = 0$ and $w^l = 0$ for $l = 1, \ldots, M$.
2: **for** Each time step **do**
3:      Obtain the inputs $\bar{x}_t$.
4:      Calculate the output of the actor u_t in (5.35).
5:      Calculate the output of the critic $\hat{V}_t$ in (5.41).
6:      Run the game for the current time step.
7:      Obtain the reward r_{t+1} and new inputs $\bar{x}_{t+1}$.
8:      Calculate $\hat{V}_{t+1}$ based on (5.41).
9:      Calculate the prediction error Δ in (5.42).
10:    Update ζ^l_{t+1} in (5.43) and w^l_{t+1} in (5.37).
11: **end for**

5.5 Homicidal Chauffeur Differential Game

DGs [2] are a family of dynamic, continuous-time games. The homicidal chauffeur DG is one type of DG. It was originally presented by Isaacs in 1954. A pursuer or a group of pursuers attempt to capture one or a group of evaders in minimal time, while the evaders try to avoid being captured.

The game terminates when the evader is within the lethal range of the pursuer (capture or termination time), or when the time exceeds 1 min (escape). Players evaluate the current state and then select their next actions. The players' strategies are not shared, and therefore each player has no knowledge of the other player's next selected action. We assume that the environment is obstacle-free.

The existence of optimal strategies in the pursuit–evasion DG is determined by Isaacs' condition [26–28]. The formal results concerning optimal strategies for pursuit–evasion DGs are given in References 26, 29. The homicidal chauffeur game and Isaacs condition for the game are discussed below.

In our model, a homicidal chauffeur game is played by autonomous robots. The chauffeur (the pursuer P) is a car-like mobile robot, and the pedestrian (the evader E) is a point that can move in any direction instantaneously. In Isaacs' homicidal chauffeur DG, a pursuer aims to minimize the capture time of an evader. The evader's objective is to maximize the capture time and avoid capture.

We assume that the players move at a constant forward speed w_i. The pursuer's speed is greater than the evader's speed, but the evader can move in any direction. The steering angle of the pursuer is given as $-u_{p_{max}} \leq u_p \leq u_{p_{max}}$, where $u_{p_{max}}$ is the maximum steering angle. The maximum steering angle results in a minimum turning radius R_p defined by

$$R_p = \frac{L_p}{\tan(u_{p_{max}})} \tag{5.45}$$

where L_p is the pursuer's wheelbase.

The dynamic equations for the pursuer P and the evader E are [26]

$$\dot{x}_p = w_p \cos(\theta_p)$$
$$\dot{y}_p = w_p \sin(\theta_p)$$
$$\dot{\theta}_p = \frac{w_p}{R_p} u_p$$
$$\dot{x}_e = w_e \cos(u_e)$$
$$\dot{y}_e = w_e \sin(u_e) \tag{5.46}$$

where (x, y), w, and θ denote the position, the velocity, and the orientation, respectively, as shown in Fig. 5-8. The angle difference ϕ between the pursuer and the evader is given as

$$\phi = \tan^{-1}\left(\frac{y_e - y_p}{x_e - x_p}\right) - \theta_p \tag{5.47}$$

The relative distance between pursuer and evader is found as

$$d = \sqrt{(x_e - x_p)^2 + (y_e - y_p)^2} \tag{5.48}$$

The capture occurs when the distance $d \leq \ell$.

In Reference ([26], pp. 232–237), Isaacs presented a condition for the pursuer to succeed in capturing the evader. Assuming that the pursuer's speed is greater than the evader's, the capture condition is given as

$$l/R_p > \sqrt{1 - \gamma^2} + \sin^{-1}\gamma - 1 \tag{5.49}$$

where ℓ/R_p is the ratio of the radius of capture to the minimum turning radius of the pursuer, and $\gamma = w_e/w_p < 1$ is the ratio of the evader's speed to the pursuer's speed. If the inequality (5.49) is reversed, E escapes from P indefinitely.

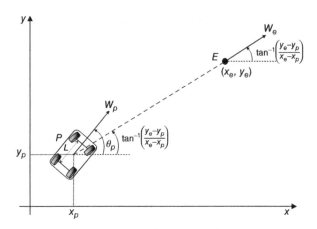

Fig. 5-8. Homicidal chauffeur problem model. Reproduced from [13], © B, Al Faiya.

Based on the capture condition in (5.49) and Isaacs' solution of the problem, the evader's optimal strategy can be obtained by solving the following two problems [26, 29, 30]:

1. When the evader is far enough from the pursuer, the evader's control strategy is to maximize the distance between the evader and the pursuer as follows:

$$u_e = \tan^{-1}\frac{y_e - y_p}{x_e - x_p} \tag{5.50}$$

2. When the distance between the pursuer and evader becomes such that $d \le R_p$, the evader adopts a second control strategy to avoid capture. The pursuer cannot turn more than a minimum turning radius R_p. The evader will make a sharp turn, normal to its direction, and enter the pursuer's non-holonomic constraint region. As shown in Fig. 5-9, a non-holonomic player is constrained to move along paths of bounded curvature such as the pursuer's minimum turning radius R_p given in Eq. (5.45). The evader's second control strategy is given as

$$u_{e_{extreme}} = \theta_e \pm \pi/2 \tag{5.51}$$

The pursuer's optimal control strategy is to minimize the distance and capture the evader in minimum time. The pursuer controls its steering angle as follows [28, 29, 31]:

$$u_p = \tan^{-1}\left(\frac{y_e - y_p}{x_e - x_p}\right) - \theta_p \tag{5.52}$$

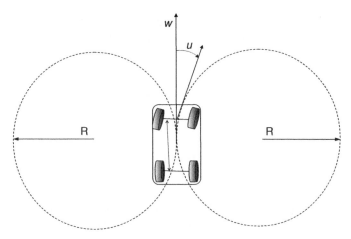

Fig. 5-9. The vehicle cannot turn into the circular region defined by its minimum turning radius *R*. Reproduced from [13], © B, Al Faiya.

5.6   Fuzzy Controller Structure

We use two inputs (fuzzy variables) to the fuzzy controller and generate one output from the fuzzy controller. The inputs for the pursuer are the angle difference ϕ and its rate of change $\dot{\phi}$. The inputs for the evader are the angle difference ϕ and the distance d. We add the distance as an input to the fuzzy controller for the evader. The reason is that the evader has higher maneuverability than the pursuer and the distance between the evader and the pursuer is critical for the evader to decide if it needs to make a sharp turn.

For simplicity and to avoid the curse of dimensionality, we use two inputs and three fuzzy sets for each input to construct the controller. The pursuer's fuzzy sets are negative (N), zero (Z), and positive (P) for the angle difference ϕ and its derivative $\dot{\phi}$. The evader's fuzzy sets are negative (N), zero (Z), and positive (P) for the angle, and far (F), close (C), and very close (V) for the distance.

We apply a zero-order TS FIS [11]. TS FIS consists of fuzzy IF-THEN rules and a fuzzy inference engine. Given the fuzzy variables x_i and the corresponding fuzzy sets A_i and B_i, the fuzzy IF-THEN rules are

$$\mathfrak{R}_l : IF\ x_1\ is\ A_l\ AND\ x_2\ is\ B_l\ THEN\ f_l = K^l \qquad (5.53)$$

where x_i represents ϕ and $\dot{\phi}$ for the pursuer, ϕ and d for the evader. f_l is the output function of rule l and K^l is the parameter for the consequence part of the fuzzy rules.

Three MFs are used for each input, which results in constructing $3^2 = 9$ rules. The Gaussian MFs are given as

$$\mu_{A_l}(x_i) = \exp\left(-\left(\frac{x_i - c_i^l}{\sigma_i^l}\right)^2\right) \tag{5.54}$$

The Gaussian MF parameters are the mean c and the standard deviation σ. We use reinforcement learning to learn the MF parameters. Figure 5-10a and b show the initial MF's before tuning.

The steering angle u is the output formed by the weighted average defuzzifier, and expressed as

$$u = \frac{\sum_{l=1}^{9}\left(\left(\prod_{i=1}^{2}\mu_{A_l}(x_i)\right)K^l\right)}{\sum_{l=1}^{9}\left(\prod_{i=1}^{2}\mu_{A_l}(x_i)\right)} \tag{5.55}$$

The fuzzy rules are illustrated using the tabular format. Tables 5.2 and 5.3 show the fuzzy decision table and the output constant K^l for the pursuer and the evader, respectively, before learning.

5.7   Q(λ)-Learning Fuzzy Inference System

A learning agent in a reinforcement learning problem interacts with the environment and receives a reward r_t at each time step t. The agent's goal is to

Table 5.2   Pursuer's fuzzy decision table before learning.

$\dot{\phi}$ / ϕ	N	Z	P
N	−0.5	−0.25	0.0
Z	−0.25	0.0	0.25
P	0.0	0.25	0.5

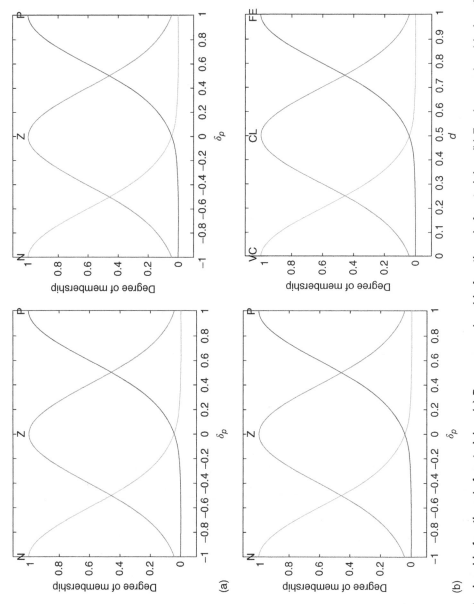

Fig. 5-10 Membership functions before training. (a) Pursuer membership functions before training. (b) Evader membership functions before training. Reproduced from [13], © B, Al Faiya.

Table 5.3 Evader's fuzzy decision table before learning.

ϕ \ d	VC	CS	FA
N	$-\pi/2$	$-\pi/2$	$-\pi/4$
Z	$-\pi/2$	$\pi/2$	0.0
P	$\pi/2$	$\pi/2$	$\pi/4$

maximize the long-run discounted return R_t [23]

$$R_t = \sum_{k=0}^{T} \gamma^k r_{t+k+1} \tag{5.56}$$

where $(0 \le \gamma \le 1)$ is the discount factor, t is the current time step, and T is the episode terminal time.

One common type of reinforcement learning is Q-learning. Q-learning estimates the action-value function $Q(s, a)$ to achieve the best expected return. The action-value function is given as

$$Q(s, a) = E\left\{ \sum_{k=0}^{\infty} \gamma^k r_{k+t+1} | s_t = s, a_t = a \right\} \tag{5.57}$$

Desouky and Schwartz [31] proposed the $Q(\lambda)$-learning fuzzy inference system (QLFIS) technique. In Reference 31, QLFIS was successfully applied to train the pursuer to capture the evader in minimum time. We apply Desouky's QLFIS algorithms for our model to train both the evader and the pursuer. The construction of the learning system is shown in Fig. 5-11. $Q(\lambda)$-learning is used to tune the input and the output parameters of the fuzzy logic controller (FLC) and the function approximator which is implemented by an FIS. In Reference 31, Desouky derived and presented the update rules of the learning process.

As shown in Fig. 5-11, the TD error δ_t is given as

$$\delta_t = r_{t+1} + \gamma \max_{\hat{a}} Q_t(s_{t+1}, \hat{a}) - Q_t(s_t, a_t) \tag{5.58}$$

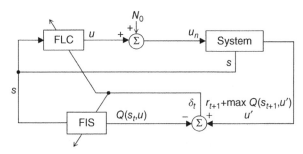

Fig. 5-11. Construction of the learning system where the white Gaussian noise $\mathcal{N}(0, \sigma_n^2)$ is added as an exploration mechanism. Reproduced from [13], © B, Al Faiya.

We apply the eligibility traces e_t to Eq. (5.57). The $Q(\lambda)$-learning action-value function in is updated by

$$Q_{t+1}(s_t, a_t) = Q_t(s_t, a_t) + \alpha \delta_t e_t \qquad (5.59)$$

where $(0 < \alpha \leq 1)$ is the learning rate, and the replacing eligibility traces is represented as

$$e_t = \gamma \lambda \, e_{t-1} + \frac{\partial Q_t(s_t, a_t)}{\partial \xi} \qquad (5.60)$$

Given the parameters $\xi = [K \ c \ \sigma]^{\mathsf{T}}$ to be tuned in the fuzzy systems, the update rules for the FIS are defined by Reference 31

$$\xi_{FIS}(t + 1) = \xi_{FIS}(t) + \eta \delta_t \left\{ \gamma \lambda e_{t-1} + \frac{\partial Q_t(s_t, u_t)}{\partial \xi_{FIS}} \right\} \qquad (5.61)$$

and the update rules for the FLC are defined by

$$\xi_{FLC}(t + 1) = \xi_{FLC}(t) + \zeta \delta_t \left\{ \frac{\partial u}{\partial \xi_{FLC}} \left(\frac{u_n - u}{\sigma_n} \right) \right\} \qquad (5.62)$$

where

$$\frac{\partial Q_t(s_t, u_t)}{\partial \xi_{FIS}} = \begin{bmatrix} \dfrac{\partial Q_t(s_t, u_t)}{\partial K^l} \\[2mm] \dfrac{\partial Q_t(s_t, u_t)}{\partial \sigma_i^l} \\[2mm] \dfrac{\partial Q_t(s_t, u_t)}{\partial c_i^l} \end{bmatrix} = \begin{bmatrix} \dfrac{\sum_l \bar{\omega}_l}{\sum_l \omega_l} \\[2mm] \dfrac{(K^l - Q_t(s_t, u_t))}{\sum_l \omega_l} \omega_l \dfrac{2(x_i - c_i^l)}{(\sigma_i^l)^2} \\[2mm] \dfrac{(K^l - Q_t(s_t, u_t))}{\sum_l \omega_l} \omega_l \dfrac{2(x_i - c_i^l)^2}{(\sigma_i^l)^3} \end{bmatrix} \qquad (5.63)$$

$$\frac{\partial u}{\partial \xi_{FLC}} = \begin{bmatrix} \dfrac{\partial u}{\partial K^l} \\[2mm] \dfrac{\partial u}{\partial \sigma_i^l} \\[2mm] \dfrac{\partial u}{\partial c_i^l} \end{bmatrix} = \begin{bmatrix} \dfrac{\sum_l \bar{\omega}_l}{\sum_l \omega_l} \\[4mm] \dfrac{(K^l - u)}{\sum_l \omega_l} \omega_l \dfrac{2(x_i - c_i^l)^2}{(\sigma_i^l)^3} \\[4mm] \dfrac{(K^l - u)}{\sum_l \omega_l} \omega_l \dfrac{2(x_i - c_i^l)}{(\sigma_i^l)^2} \end{bmatrix} \tag{5.64}$$

with the learning rate η for the FIS and ζ for the FLC. The firing strength ω_l and the normalized firing strength $\bar{\omega}_l$ of the rule l are defined as follows [31]:

$$\omega_l = \prod_{i=1}^{2} \exp\left(-\left(\frac{x_i - c_i^l}{\sigma_i^l}\right)^2\right) \tag{5.65}$$

$$\bar{\omega}_l = \frac{\omega_l}{\sum_{l=1}^{9} \omega_l} \tag{5.66}$$

The learning algorithm used in our simulation is shown in Algorithm 5.3, where M is the number of episodes (games) and N is the number of steps (plays).

Algorithm 5.3 QLFIS algorithm

1: MFs ← Fig. 5.10
2: K^l ← Tables 5.2 and 5.3
3: $Q(s, u) \leftarrow 0$ {FIS Q-values}
4: $e \leftarrow 0$ {eligibility traces of the FIS}
5: $\gamma \leftarrow 0.95$; $\lambda \leftarrow 0.9$; $\sigma_n \leftarrow 0.08$
6: **for** $i \leftarrow 1$ **to** M **do**
7:    $\eta \leftarrow \left(0.1 - 0.09\left(\dfrac{i}{M}\right)\right)$
8:    $\zeta \leftarrow \left(0.01 - 0.009\left(\dfrac{i}{M}\right)\right)$
9:    $(x_p, y_p) \leftarrow (0, 0)$ {pursuer initial position}
10:   initialize (x_e, y_e) randomly   {evader initial position}
11:   update $s_p = (\phi, \dot{\phi})$
12:   update $s_e = (\phi, d)$
13:   $u \leftarrow$ Eq. (5.55)   {for the pursuer and the evader}
14:   **for** $j \leftarrow 1$ **to** N **do**
15:     $u_n \leftarrow u + \mathcal{N}_0$ {for the pursuer and the evader}
16:     $Q(s_t, u) \leftarrow$ Eq. (5.59)

17:    play the game, observe the next states s'_p and s'_e and the reward r
18:    $Q_{(s_{t+1}, u')} \leftarrow$ Eq. (5.55)
19:    $\delta_t \leftarrow$ Eq. (5.58)
20:    $e_t \leftarrow$ Eq. (5.60)   {for the FIS}
21:    $\xi(t+1)_{FIS} \leftarrow$ Eq. (5.61)   {update FIS input and output parameters}
22:    $\xi(t+1)_{FLC} \leftarrow$ Eq. (5.62)   {update FLC input and output parameters}
23:    $s_t \leftarrow s_{t+1}$; $u \leftarrow u'$
24:    **end for**
25: **end for**

5.8   Simulation Results for the Homicidal Chauffeur

We simulate the system for different numbers of episodes. The number of steps is 600, and the sampling time is 0.1 s.

The pursuer is twice as fast as the evader such that $w_p = 2$ m/s and $w_e = 1$ m/s. The pursuers wheelbase is $L = 0.3$ m. At each episode, the position of the evader is initialized randomly. The initial position of the pursuer is at the origin $(x_p, y_p) = (0,0)$, and the initial orientation of the pursuer is $\theta = 0$ rad. We consider the kinematic equations of the pursuer and the evader given in Eq. (5.46).

The game terminates when the pursuer captures the evader, or when the time exceeds 60 s (escape). The capture radius is $\ell \leq 0.15$ m which is half the wheelbase of the pursuer $\ell \leq \frac{L_p}{2}$. The maximum steering angle of the pursuer is $-0.5 \leq u_{p_{max}} \leq 0.5$ rad with $R_p = 0.5491$ m. Given the stated parameters of the system and using Isaacs' capture condition, there exists a strategy for the evader to avoid capture.

We now use the learning algorithm presented in Algorithm 5.3 to simulate the game. The initial conditions of the learning system are given in steps 1–5 of Algorithm 5.3. To evaluate the learning technique efficiency of the evader, the pursuer's parameters are initialized such that the pursuer can capture the evader with their initial strategies before learning. We apply the same learning algorithm QLFIS to both players.

At the beginning of the learning, the pursuer always captured the evader, as shown in Fig. 5-12. After 500 episodes, in Fig. 5-13, the evader increased the capture time and made successful maneuvers. Figure 5-14 and Table 5.4 show that the evader learned to escape from the pursuer after 1000 episodes. The

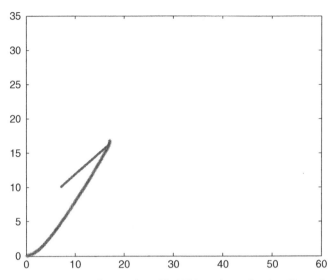

Fig. 5-12. The pursuer captures the evader with 100 learning episodes. Reproduced from [13], © B, Al Faiya.

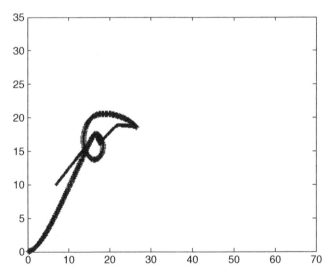

Fig. 5-13. The evader increases the capture time after 500 learning episodes. Reproduced from [13], © B, Al Faiya.

evader makes sharp turns when the distance $d \leq R_p$. The evader avoids capture by changing its direction and entering into the pursuer's turning radius constraint. The learners' fuzzy consequence parameters K^l after 1000 episodes are shown in Tables 5.5 and 5.6.

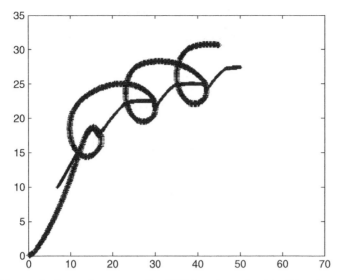

Fig. 5-14. The evader learns to escape after 1000 learning episodes. Reproduced from [13], © B, Al Faiya.

Table 5.4   Capture time (s) for different numbers of learning episodes.

Game	No. of episodes	Capture time (s)
Theoretical	—	>60 (escape)
After learning using QLFIS	100	12.90
	500	25.10
	1000	>60 (escape)

Table 5.5   The evader's fuzzy decision table after 1000 learning episodes.

ϕ \ d	VC	CL	FE
N	−1.5848	−1.5782	−0.4074
Z	−1.5758	1.5526	0.0331
P	1.5930	1.5794	0.2626

Table 5.6   The pursuer's fuzzy decision table after 1000 learning episodes.

$\dot{\phi}$ ϕ	N	Z	P
N	−0.4763	−0.2503	−0.0075
Z	−0.2413	0.0023	0.1522
P	−0.0046	0.2650	0.4777

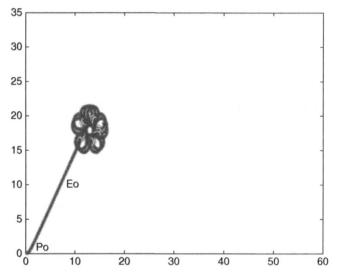

Fig. 5-15. The evader avoids capture when $u_{p_{max}}$ = 0.5 rad. Reproduced from [13], © B, Al Faiya.

For comparison, we show the results of the theoretical solution described in Section 5.5. Given the stated parameters of the system and using Isaacs capture condition, there exists a strategy for the evader to avoid capture when $-0.5 \leq u_{p_{max}} \leq 0.5$ rad. Figure 5-15 shows that the evader can escape from the pursuer by making sharp turns. We then increase the maximum steering angle of the pursuer to $-0.7 \leq u_{p_{max}} \leq 0.7$ rad. In this case, the capture condition is satisfied. The pursuer can capture the evader as shown in Fig. 5-16 with capture time = 11.90 s.

5.9   Learning in the Evader–Pursuer Game with Two Cars

The pursuit–evasion model is shown in Fig. 5-17. Equations of motion for the pursuer/evader robot are [29]

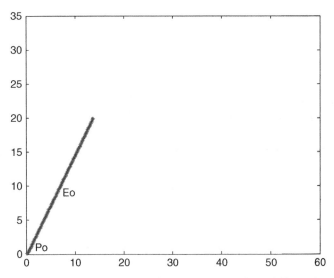

Fig. 5-16. The pursuer can capture the evader when $u_{p_{max}} = 0.7$ rad. Reproduced from [13], © B, Al Faiya.

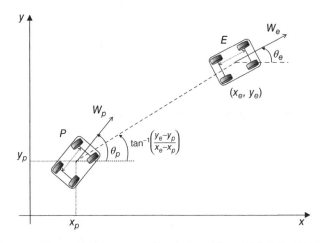

Fig. 5-17. The game of two cars. Reproduced from [13], © B, Al Faiya.

$$\dot{x}_i = v_i \cos(\theta_i)$$

$$\dot{y}_i = v_i \sin(\theta_i)$$

$$\dot{\theta}_i = \frac{v_i}{L_i} \tan(u_i) \tag{5.67}$$

where "i" is "p" for the pursuer and is "e" for the evader, (x_i, y_i) is the position of the robot, θ_i is the orientation, L_i is the robot wheelbase, u_i is the steering angle, $u_i \in [-u_{i_{max}}, u_{i_{max}}]$, and v_i is the velocity of the robot which is governed by the steering angle, to avoid slips, such that

$$v_i = V_{i_{max}} \cos(u_i) \tag{5.68}$$

where $V_{i_{max}}$ is the maximum velocity of the robot. Our scenario is to make the pursuer faster than the evader, that is, $V_{p_{max}} > V_{e_{max}}$, but at the same time to make the pursuer less maneuverable than the evader, that is, $u_{p_{max}} < u_{e_{max}}$. The control strategy for the pursuer is to drive the angle difference between the pursuer and the evader, δ, to be zero where the error in angle, δ, is calculated as

$$\delta = \tan^{-1} \left(\frac{y_e - y_p}{x_e - x_p} \right) - \theta_p \tag{5.69}$$

The control strategy for the evader is to maximize the distance between them in two ways [29, 32]:

1. If the distance between the pursuer and the evader is greater than a certain value d, then the control strategy for the evader is

$$u_e = \tan^{-1} \left(\frac{y_e - y_p}{x_e - x_p} \right) - \theta_e \tag{5.70}$$

2. If the distance between the pursuer and the evader is smaller than d, then the evader exploits its higher maneuverability. Therefore, the control strategy for the evader will be

$$u_e = (\theta_p + \pi) - \theta_e. \tag{5.71}$$

This strategy will increase the maneuverability of the evader and make it more difficult (but not impossible) for the pursuer to catch the evader. We choose this strategy to show the effect of our proposed techniques on the system. The classical control strategy of the pursuer that we compare our results with is defined by

$$u_p = \begin{cases} -u_{p_{max}} & : \quad \delta < -u_{p_{max}} \\ \delta & : \quad -u_{p_{max}} \leq \delta \leq u_{p_{max}} \\ u_{p_{max}} & : \quad \delta > u_{p_{max}} \end{cases} \tag{5.72}$$

where δ is defined by (5.69).

The capture occurs when the distance between the pursuer and the evader is less than a certain value, ℓ. We call this the capture radius, which is defined as

$$\ell = \sqrt{(x_e - x_p)^2 + (y_e - y_p)^2} \tag{5.73}$$

5.10   Simulation of the Game of Two Cars

The initial position of the pursuer is at the origin $(x_p, y_p) = (0, 0)$, and the position of the evader is initialized randomly at each episode. The initial orientation of the pursuer and the initial positions are the same as in the homicidal chauffeur game. The initial orientation of the evader is $\theta_e = 0$ rad. We use the kinematic equations of the pursuer and the evader given in (5.67).

Similarly, the game is initialized such that the pursuer can capture the evader as shown in Fig. 5-18. After 500 episodes, in Fig. 5-19, the evader increased the capture time and made a successful maneuver. Figure 5-20 and Table 5.7 show that the evader learned to escape from the pursuer after approximately 1300 episodes. The evader makes sharp turns to enter into the pursuer's turning radius constraint when the distance $d \leq R_p$.

Figures 5-21 and 5-22 show the MFs of the evader and the pursuer after learning. The consequence parameters K^l after training are shown in Tables 5.8 and 5.9.

Without Eligibility Traces. We now apply Q-learning without the use of eligibility traces. In each episode, we record the time of capture and plot the time versus 500 episodes. Then we run 10 simulations and average the result. The solid

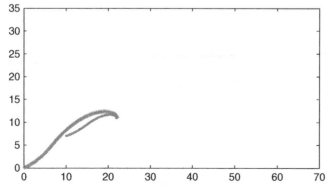

Fig. 5-18. The pursuer captures the evader with 100 learning episodes. Reproduced from [13], © B, Al Faiya.

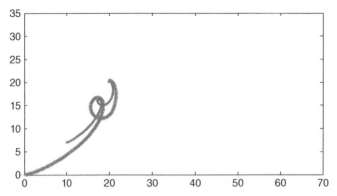

Fig. 5-19. The evader increases the capture time after 500 learning episodes. Reproduced from [13], © B, Al Faiya.

Table 5.7   Summary of the time of capture for different numbers of learning episodes in the game of two cars.

Game	No. of episodes	Capture time T_c (sec)
Theoretical	—	>60 (escape)
After learning using QLFIS	100	13.70
	500	27.50
	1300	>60 (escape)

Table 5.8   The evader's fuzzy decision table and the output constant K^l after learning.

d ϕ	V	C	F
N	−1.591	−1.572	−0.337
Z	−1.613	1.571	0.146
P	1.537	1.573	0.429

Table 5.9   The pursuer's fuzzy decision table and the output constant K^l after learning.

$\dot{\phi}$ ϕ	N	Z	P
N	−0.4660	−0.2512	−0.0005
Z	−0.3507	0.0274	0.1765
P	−0.0124	0.2615	0.4830

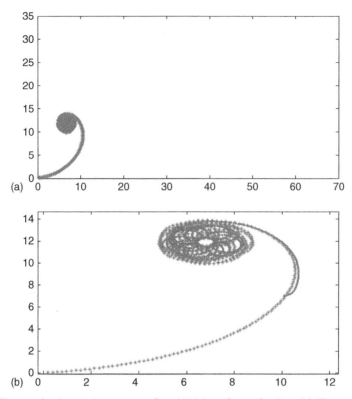

Fig. 5-20. The evader learns to escape after 1300 learning episodes. (a) The evader learns to escape after 1300 learning episodes. (b) Zoomed version of (a). Reproduced from [13], © B, Al Faiya.

line in Fig. 5-23 is the result from using $Q(\lambda)$-learning. The dashed line is the result from using Q-learning. Compared to Q-learning, the learning speed is similar to that with the eligibility traces. The convergence speed of the player's learning process has not improved significantly when using $Q(\lambda)$-learning.

This section presented the applications of fuzzy $Q(\lambda)$-learning and fuzzy Q-learning to pursuit–evasion DGs. The fuzzy controller, the convergence of the learning process, and the learning speed were investigated. The QLFIS technique was then used to train both the evader and the pursuer simultaneously. The trained evader learned to make sharp turns (extreme strategy) to maximize the time of capture and, if possible, avoid being captured. Simulation results of the homicidal chauffeur game and the game of two cars showed that the evader successfully learned to escape from the pursuer. The use of eligibility traces did not significantly improve the learning speed when used in $Q(\lambda)$-learning. Moreover, eligibility traces required more computations per episode.

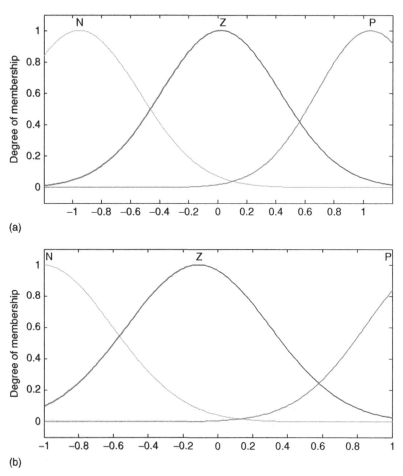

(a)

(b)

Fig. 5-21. The pursuer's membership functions after training. (a) The angle difference ϕ. (b) The rate of change of the angle difference $\dot{\phi}$. Reproduced from [13], © B, Al Faiya.

5.11   Differential Game of Guarding a Territory

In this section, we apply fuzzy reinforcement learning algorithms to the DG of guarding a territory and let the defender learn its Nash equilibrium strategy by playing against the invader. To speed up the defender's learning process, we design a shaping reward function for the defender in the game. Moreover, we apply the same FACL algorithm and shaping reward function to a three-player DG of guarding a territory including two defenders and one invader. We run simulations to test the learning performance of the defenders in both cases.

In reinforcement learning, a reinforcement learner may suffer from the temporal credit assignment problem where a player's reward is delayed or

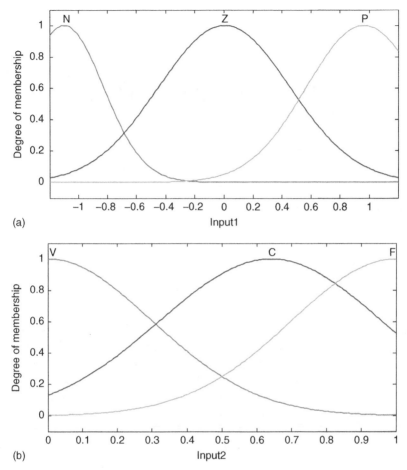

Fig. 5-22. The evader's membership functions after training. (a) The angle difference ϕ. (b) The distance between the pursuer and the evader d. Reproduced from [13], © B, Al Faiya.

only received at the end of an episodic game. When a task has a very large state space or continuous state space, the delayed reward will slow down the learning dramatically. For the game of guarding a territory, the only reward received during the game is the distance between the invader and the territory at the end of the game. Therefore, it is extremely difficult for a player to learn its optimal strategy based on this very delayed reward. To deal with the temporal credit assignment problem and speed up the learning process, one can apply reward shaping to the learning problem [33]. Shaping can be implemented in reinforcement learning by designing intermediate shaping rewards as an informative reinforcement signal to the learning agent and reward the agent for making a good estimate of the desired behavior [23, 34,

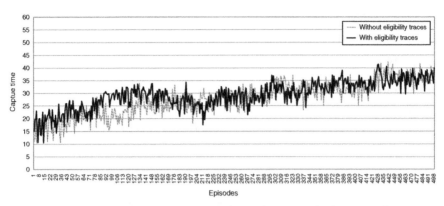

Fig. 5-23. The time of capture with the use of eligibility traces in the game of two cars. Reproduced from [13], © B, Al Faiya.

35]. The idea of reward shaping is to provide an additional reward as a hint, based on the knowledge of the problem, to improve the performance of the agent [33].

The Isaacs' guarding a territory game is a two-player zero-sum DG. The invader's goal is to reach the territory. If the invader cannot reach the territory, it at least moves to a point as close to the territory as possible [2]. Accordingly, the defender tries to intercept the invader at a point as far away as possible from the territory [2]. We denote the invader as I and the defender as D in Fig. 5-24. The dynamics of the invader I and the defender D are defined as

$$\dot{x}_D(t) = \sin\theta_D, \quad \dot{y}_D(t) = \cos\theta_D \tag{5.74}$$

$$\dot{x}_I(t) = \sin\theta_I, \quad \dot{y}_I(t) = \cos\theta_I \tag{5.75}$$

$$-\pi \leq \theta_D \leq \pi, \quad -\pi \leq \theta_I \leq \pi$$

where θ_D is the defender's strategy and θ_I is the invader's strategy.

In order to simplify the problem, we establish a relative coordinate frame centered at the defender's position with its y'-axis in the direction of the invader's position as shown in Fig. 5-24. The territory is represented by a circle with center $T(x'_T, y'_T)$ and radius R. Different from θ_D and θ_I in the original coordinate frame, we define u_D as the defender's strategy and u_I as the invader's strategy in relative coordinates.

The payoff for this game is defined as

$$P_{ip}(u_D, u_I) = \sqrt{(x'_I(t_f) - x'_T)^2 + (y'_I(t_f) - y'_T)^2} - R \tag{5.76}$$

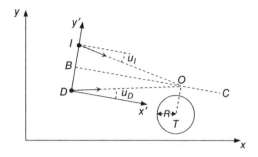

Fig. 5-24. The differential game of guarding a territory. Reproduced from [18], © X. Lu.

where ip denotes the players' initial positions, R is the radius of the target, and t_f is the terminal time. The terminal time is the time when the invader reaches the territory, or the invader is intercepted before it reaches the territory. The above payoff indicates how close the invader can move to the territory if both players start from their initial positions and follow their stationary strategies u_D and u_I thereafter. In this game, the invader tries to minimize the payoff P while the defender tries to maximize it.

In Fig. 5-24, we draw the bisector BC of the segment ID. According to the dynamics of the players in (5.74) and (5.75), the players can move in any direction instantaneously with the same speed. Therefore, the region above the line BC is where the invader can reach before the defender, and the region below the line BC is where the defender can reach before the invader. We draw a perpendicular line TO to the bisector BC through the point T. Then point O is the closest point on the line BC to the territory T. Starting from the initial position (I, D), if both players play their optimal strategies, the invader can only reach point O as its closest point to the territory.

The value of the game can be found as the shortest distance between the line BC and the territory. We define the value of the game as

$$P(u_D^*, u_I^*) = \| \overrightarrow{TO} \| - R \tag{5.77}$$

where u_D^* and u_I^* are the players' Nash equilibrium strategies given by

$$u_D^* = \angle \overrightarrow{DO}, \tag{5.78}$$

$$u_I^* = \angle \overrightarrow{IO}, \tag{5.79}$$

$$-\pi \leq u_D^* \leq \pi, -\pi \leq u_I^* \leq \pi$$

5.12   Reward Shaping in the Differential Game of Guarding a Territory

In reinforcement learning, the player may suffer from the temporal credit assignment problem where a reward is received only after a sequence of actions. For example, players in a soccer game are given rewards only after a goal is scored. This will lead to the difficulty in distributing credit or punishment to each action from a long sequence of actions. We define the terminal reward when the reward is received only at the terminal time. If the reinforcement learning problem is in the continuous domain with only a terminal reward, it is almost impossible for the player to learn without any information other than this terminal reward.

In the DG of guarding a territory, the reward is received only when the invader reaches the territory or is intercepted by the defender. According to the payoff function given in (5.76), the terminal reward for the defender is defined as

$$R_D = \begin{cases} Dist_{IT} & \text{the defender captures the invader} \\ 0 & \text{the invader reaches the territory} \end{cases} \qquad (5.80)$$

where $Dist_{IT}$ is the distance between the invader and the territory at the terminal time. Since we only have terminal rewards in the game, the learning process of the defender will be prohibitively slow. To solve this, one can use a shaping reward function for the defender to compensate for the lack of immediate rewards.

The purpose of reward shaping is to improve the learning performance of the player by providing an additional reward to the learning process. But the question is how to design good shaping reward functions for different types of games. In the pursuit–evasion game, the immediate reward is defined as

$$r_{t+1} = Dist_{ID}(t) - Dist_{ID}(t+1) \qquad (5.81)$$

where $Dist_{ID}(t)$ denotes the distance between the pursuer and the evader at time t. One might consider the above immediate reward as the shaping reward function for the DG of guarding a territory. However, the immediate reward in (5.81) is not a good candidate for the shaping reward function in our game. The goal of the pursuer is to minimize the distance between the pursuer and the evader at each time step. Different from the pursuer, the goal of the defender in the DG of guarding a territory is to keep the invader away from the territory. Since the defender and the invader have the same speed, the defender may fail to protect the territory if the defender keeps chasing after the invader all the time.

Based on the above analysis and the characteristics of the game, we design the following shaping reward function for the defender:

$$r_{t+1} = y'_T(t) - y'_T(t+1) \tag{5.82}$$

where $y'_T(t)$ and $y'_T(t+1)$ denote the territory's relative position of the y'-axis at time t and $t+1$, respectively.

The shaping reward function in (5.82) is designed based on the idea that the defender tries to protect the territory from invasion by keeping the territory and the invader on opposite sides. In other words, if the invader is on the defender's left side, then the defender needs to move in a direction where it can keep the territory as far to the right side as possible. As shown in the relative coordinates in Fig. 5-24, the invader is located on the positive side of the y'-axis. Then, the goal of the defender in Fig. 5-24 is to keep the invader on the positive side of the y'-axis and move in a direction where it can keep the territory further to the negative side of the y'-axis.

5.13   Simulation Results

We assume that the defender does not have any information about its optimal strategy or the invader's strategy. The only information the defender has is the players' current positions. We apply the aforementioned FQL and FACL algorithms in Section 5.4 to the game and make the defender learn to intercept the invader. To compensate for the lack of immediate rewards, the shaping reward functions introduced in Section 5.12 are added to the FQL and FACL algorithms. Simulations are conducted to show the learning performance of the FQL and FACL algorithms based on different reward functions. Then we add one more defender to the game. We use the same FACL algorithm to both defenders independently. Each defender only has its own position and the invader's position as the input signals. Then the FACL algorithm becomes a completely decentralized learning algorithm in this case. We test, through simulations, how the two defenders can cooperate with each other to achieve good performance even though they do not directly share any information.

5.13.1   One Defender Versus One Invader

We first simulate the DG of guarding a territory with two players whose dynamics are given in (5.74) and (5.75).

To reduce the computational load, $\mu^{F_i^l}(x_i)$ in (5.26) is defined as a triangular MF. In this game, we define three input variables, which are the relative y-position y'_I of the invader, the relative x-position x'_T of the territory, and the

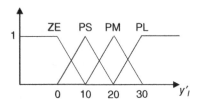

Fig. 5-25. MFs for y'_l. Reproduced from [18], with permission of Carleton University.

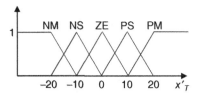

Fig. 5-26. Membership functions for input variables. Reproduced from [18], with permission of Carleton University.

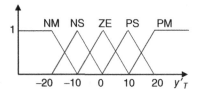

Fig. 5-27. Membership functions for input variables. Reproduced from [18], © X. Lu.

relative y-position y'_T of the territory. The predefined triangular MFs for each input variable are shown in Figs. 5-25–5-27. The number of fuzzy rules applied to this game is $4 \times 5 \times 5 = 100$. The selection of the number of rules and the MFs in the premise part of the fuzzy rules is based on the prior knowledge of the game.

For the FQL algorithm, we pick the discrete action set A as

$$A = \{\pi, 3\pi/4, \pi/2, \pi/4, 0, -\pi/4, -\pi/2, -3\pi/4\} \qquad (5.83)$$

The ε-greedy policy in (5.29) is set to $\varepsilon = 0.2$. For the FACL algorithm, we set the learning rate $\alpha = 0.1$ in (5.43) and $\beta = 0.05$ in (5.36). The exploration policy in the FACL algorithm is chosen as a random white noise $v(0, \sigma)$ with $\sigma = 1$. To reduce the influence of the future rewards to the current state, we choose a small discount factor $\gamma = 0.5$ in (5.33) and (5.42).

We now define episodes and training trials for the learning process. An *episode* or a single run of the game is when the game starts at the players' initial positions and ends at a terminal state. A *terminal state* in this game is the state where the defender captures the invader or the invader enters the territory. A *training trial* is defined as one complete learning cycle which contains 200 training episodes. We set the invader's initial position at $(5, 25)$ for each training episode. The center of the territory is located at $(20, 10)$ with radius $R = 2$.

Example 5.2   We assume that the invader plays its Nash equilibrium strategy all the time. The defender, starting at the initial position $(5, 5)$, learns to intercept the NE invader. We call the invader that always plays its Nash equilibrium strategy as the *NE invader*. We run simulations to test the performance of the FQL and FACL algorithms with different shaping reward functions introduced in Section 5.12. Figures 5-28–5-30 show the simulation results after one training trial including 200 training episodes. In Fig. 5-28, with only the terminal reward given in (5.80), the trained defender failed to intercept the invader. The same happened when the shaping reward function given in (5.81) was used in the FQL and the FACL algorithms, as shown in Fig. 5-29. As we discussed in Section 5.12, the shaping reward function in (5.81) is not a good candidate for this game. With the help of our proposed shaping reward function in (5.82), the trained defender successfully intercepted the invader, as shown in Fig. 5-30. This example verifies the importance of choosing a good shaping reward function for the FQL and FACL algorithms for our game.

Example 5.3   In this example, we show the average performance of the FQL and FACL algorithms with the proposed shaping reward function given in (5.82).

The training process includes 20 training trials with 200 training episodes for each training trial. For each training episode, the defender randomly chooses one initial position from the defender's initial positions 1–4 shown in Fig. 5-31. After every 10 training episodes in each training trial, we set up a testing phase to test the performance of the defender trained so far. In the testing phase, we let the NE invader play against the trained defender and calculate the performance error as follows:

$$PE_{ip} = P_{ip}(u_D^*, u_I^*) - P_{ip}(u_D, u_I^*), \qquad (ip = 1, \dots, 6) \qquad (5.84)$$

where ip represents the initial positions of the players, the payoffs $P_{ip}(u_D^*, u_I^*)$ and $P_{ip}(u_D, u_I^*)$ are calculated based on (5.76), and PE_{ip} denotes the calculated performance difference for players' initial positions ip. In this example, the invader's initial position is fixed during learning. Therefore, the players' initial positions ip are represented as the defender's initial positions 1–6

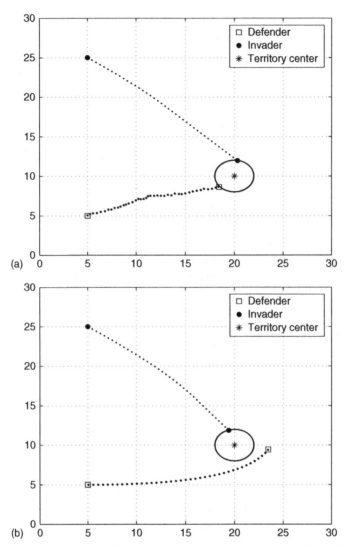

Fig. 5-28. Reinforcement learning with no shaping function in Example 5.2. (a) Trained defender using FQL with no shaping function. (b) Trained defender using FACL with no shaping function. Reproduced from [18], © X. Lu.

shown in Fig. 5-31. We use $PE_{ip}(TE)$ to represent the calculated performance error for the defender's initial position ip at the TEth training episode. For example, $PE_1(10)$ denotes the performance error calculated based on (5.84) for defender's initial position 1 at the 10th training episode. Then we average the performance error over 20 trails and get

$$\overline{PE}_{ip}(TE) = \frac{1}{20} \sum_{Trl=1}^{20} PE_{ip}^{Trl}(TE), \qquad (ip = 1, \dots, 6) \qquad (5.85)$$

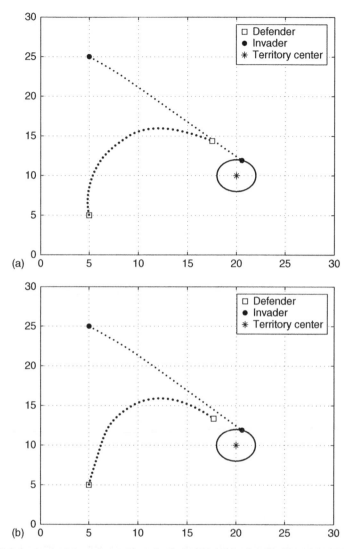

Fig. 5-29. Reinforcement learning with a bad shaping function in Example 5.2. (a) Trained defender using FQL with the bad shaping function in Example 5.2. (b) Trained defender using FACL with the bad shaping function in Example 5.2. Reproduced from [18], © X. Lu.

where $\overline{PE}_{ip}(TE)$ denotes the averaged performance error for players' initial position ip at the TEth training episode over 20 training trails.

Figure 5-32 shows the results where the average performance error $\overline{PE}_{ip}(TE)$ becomes smaller after learning for the FQL and the FACL algorithms. Note that the defender's initial positions 5 and 6 in Fig. 5-31 are not included in the training episodes. Although we did not train the defender's initial positions 5

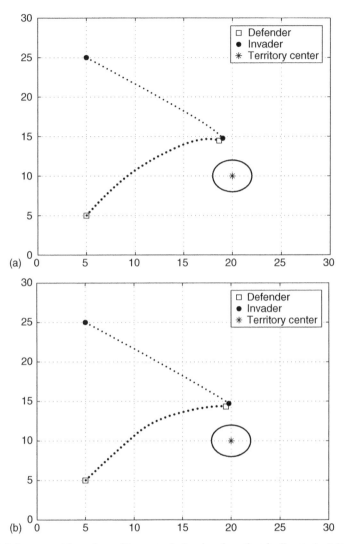

Fig. 5-30. Reinforcement learning with a good shaping function in Example 5.2. (a) Trained defender using FQL with the good shaping function in Example 5.2. (b) Trained defender using FACL with the good shaping function in Example 5.2. Reproduced from [18], with permission of Carleton University.

and 6, the convergence of the performance errors PE_5 and PE_6 verify that the defender's learned strategy is close to its NE strategy. Compared to Fig. 5-32a for the FQL algorithm, in Fig. 5-32b the performance errors for the FACL algorithm converge closer to zero after the learning. The reason is that the global continuous action in (5.30) for the FQL algorithm is generated based on a fixed discrete action set A with only eight elements given in (5.83). The closeness of the defender's learned action (strategy) to its NE action (strategy)

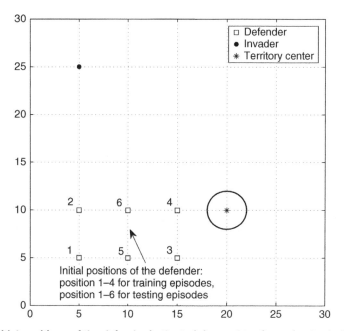

Fig. 5-31. Initial positions of the defender in the training and testing episodes in Example 5.3. Reproduced from [18], with permission of Carleton University.

is determined by the size of the action set A in the FQL algorithm. A larger size of the action set encourages the convergence of the defender's action (strategy) to its NE action (strategy), but the increasing dimension of the Q function will cause a slow learning speed, as we discussed at the beginning of Section 5.4. For the FACL algorithm, the defender's global continuous action is updated directly by the prediction error in (5.42). In this way, the convergence of the defender's action (strategy) to its NE action (strategy) is better in the FACL algorithm.

5.13.2   Two Defenders Versus One Invader

We now add a second defender to the game with the same dynamics as the first defender as defined in (5.74). The payoff for this game is defined as

$$P(u_{D1}, u_{D2}, u_I) = \sqrt{(x_I'(t_f) - x_T')^2 + (y_I'(t_f) - y_T')^2} - R \qquad (5.86)$$

where u_{D_1}, u_{D_2}, and u_I are the strategies for defender 1, defender 2, and the invader, respectively, and R is the radius of the target. Based on the analysis of the two-player game in Section 5.11, we can also find the value of the game for the three-player DG of guarding a territory. For example, we call the

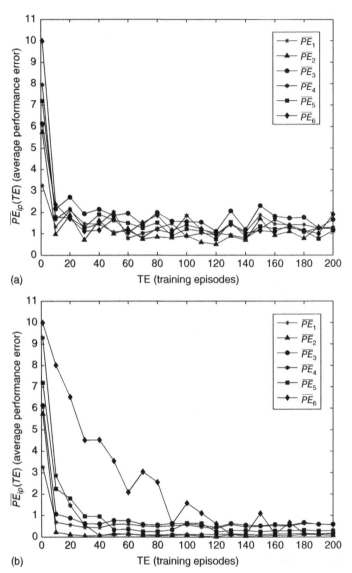

Fig. 5-32. Example 5.3: average performance of the trained defender versus the NE invader. (a) Average performance error $\overline{PE}_{ip}(TE)(ip = 1, \ldots, 6)$ in the FQL algorithm. (b) Average performance error $\overline{PE}_{ip}(TE)(ip = 1, \ldots, 6)$ in the FACL algorithm. Reproduced from [18], © X. Lu.

gray region in Fig. 5-33 as the invader's reachable region where the invader can reach before the two defenders. Then the value of the game becomes the shortest distance from the territory to the invader's reachable region. In Fig. 5-33, point O on the invader's reachable region is the closest point to the

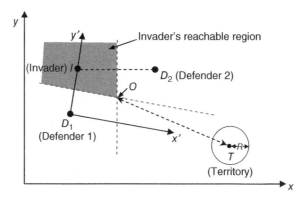

Fig. 5-33. The differential game of guarding a territory with three players. Reproduced from [18], © X. Lu.

territory. Therefore, the value of the game becomes

$$P(u_{D_1}^*, u_{D_2}^*, u_I^*) = \parallel \overrightarrow{TO} \parallel - R \tag{5.87}$$

where $u_{D_1}^*, u_{D_2}^*, u_I^*$ are the NE strategies for defender 1, defender 2, and the invader, respectively. Based on (5.87), the players' NE strategies are given as

$$u_{D_1}^* = \angle \overrightarrow{D_1 O}, \tag{5.88}$$

$$u_{D_2}^* = \angle \overrightarrow{D_2 O}, \tag{5.89}$$

$$u_I^* = \angle \overrightarrow{IO}, \tag{5.90}$$

$$-\pi \le u_{D_1}^* \le \pi, -\pi \le u_{D_2}^* \le \pi, -\pi \le u_I^* \le \pi$$

We apply the FACL algorithm to the game and make the two defenders learn to cooperate to intercept the invader. The initial position of the invader and the position of the target are the same as in the two-player game. Each defender in this game uses the same parameter settings of the FACL algorithm as in Section 5.13.1. Moreover, each defender has the information of only its own position and the invader's position without any information from the other defender. Each defender uses the same FACL algorithm independently, which makes the FACL algorithm a completely decentralized learning algorithm in this game.

Example 5.4　We assume that the invader plays its Nash equilibrium strategy given in (5.90) all the time. The two defenders, starting at the initial position (5, 5) for defender 1 and (25, 25) for defender 2, learn to intercept the NE

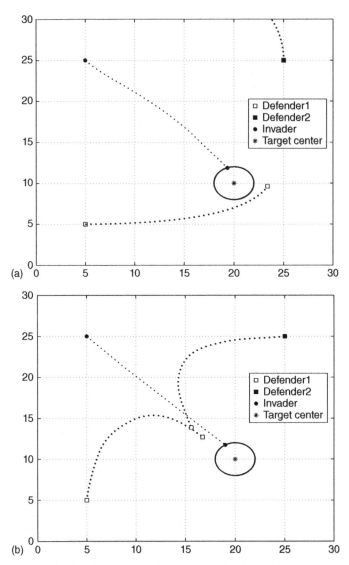

Fig. 5-34. Reinforcement learning without shaping or with a bad shaping function in Example 5.4. (a) Two trained defenders using FACL with no shaping function versus the NE invader after one training trial. (b) Two trained defenders using FACL with the bad shaping function versus the NE invader after one training trial. Reproduced from [18], © X. Lu.

invader. Similar to the two-player game in Section 5.13.1, we run a single trial including 200 training episodes to test the performance of the FACL algorithm with different shaping reward functions given in Section 5.12. In Fig. 5-34, two defenders failed to intercept the NE invader with only the terminal reward

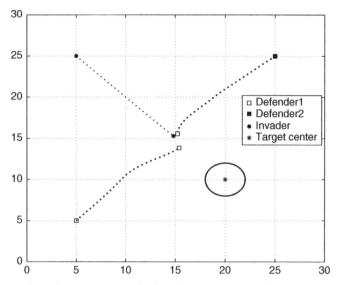

Fig. 5-35. Two trained defenders using FACL with the good shaping function versus the NE invader after one training trial in Example 5.4. Reproduced from [18], © X. Lu.

and with the shaping reward function given in (5.81). On the contrary, with the proposed shaping reward function in (5.82), the two trained defenders successfully intercepted the NE evader after one training trial, as shown in Fig. 5-35.

Example 5.5 In this example, we show the average performance of the FACL algorithm with our proposed shaping reward function for the three-player game. Similar to Example 5.3, we run 20 training trials with 200 training episodes for each training trial. The training process includes 20 training trials with 200 training episodes for each training trial. For each training episode, the defender randomly chooses one initial position from the defender's initial positions 1–2 shown in Fig. 5-36a.

After every 10 training episodes, we set up a testing phase to test the performance of the defender trained so far. The performance error in a testing phase is defined as

$$PE_{ip} = P_{ip}(u_{D_1}^*, u_{D_2}^*, u_I^*) - P_{ip}(u_{D_1}, u_{D_2}, u_I^*), \qquad (ip = 1, \dots, 4) \qquad (5.91)$$

where ip represents the defender's initial positions 1–4 shown in Fig. 5-36a, and $P_{ip}(u_{D_1}^*, u_{D_2}^*, u_I^*)$ and $P_{ip}(u_{D_1}, u_{D_2}, u_I^*)$ are the payoffs calculated based on

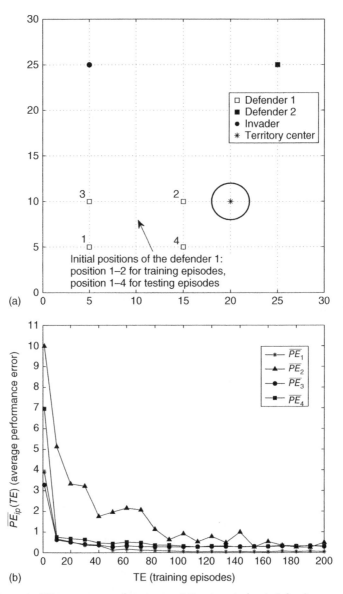

Fig. 5-36. Example 5.5: average performance of the two trained defenders versus the NE invader. (a) Initial positions of the players in the training and testing episodes. (b) Average performance error for the trained defenders versus the NE invader. Reproduced from [18], © X. Lu.

(5.86). Then we average the performance error over 20 trails and get

$$\overline{PE}_{ip}(TE) = \frac{1}{20} \sum_{Trl=1}^{20} PE_{ip}^{Trl}(TE), \qquad (ip = 1, \dots, 4) \qquad (5.92)$$

where $\overline{PE}_{ip}(TE)$ denotes the averaged performance error for players' initial position ip at the TEth training episode over 20 training trails. The simulation result in Fig. 5-36b shows that the average performance error $\overline{PE}_{ip}(TE)$ ($ip = 1, \ldots, 4$) converges close to zero after 200 training episodes. Based on the simulation results, the two trained defenders successfully learned to intercept the NE invader. Although there is no training performed for positions 3 and 4, as shown in Fig. 5-36a, the convergence of PE_3 and PE_4 in Fig. 5-36b verifies the good performance of two trained defenders. Simulation results also verify the effectiveness of the proposed shaping reward function to the FACL algorithm in the three-player DG of guarding a territory.

References

[1] R. Abielmona, E. Petriu, M. Harb, and S. Wesolkowski, "Mission-driven robotic intelligent sensor agents for territorial security," *IEEE Computational Intelligence Magazine*, vol. 6, no. 1, pp. 55–67, 2011.

[2] R. Isaacs, *Differential Games: A Mathematical Theory with Applications to Warfare and Pursuit, Control and Optimization*. New York: John Wiley and Sons, Inc., 1965.

[3] M. E. Harmon, L. C. Baird III, and A. H. Klopf, "Reinforcement learning applied to a differential game," *Adaptive Behavior*, vol. 4, no. 1, pp. 3–28, 1995.

[4] J. W. Sheppard, "Colearning in differential games," *Machine Learning*, vol. 33, pp. 201–233, 1998.

[5] S. N. Givigi, H. M. Schwartz, and X. Lu, "A reinforcement learning adaptive fuzzy controller for differential games," *Journal of Intelligent and Robotic Systems*, vol. 59, pp. 3–30, 2010.

[6] S. F. Desouky and H. M. Schwartz, "Self-learning fuzzy logic controllers for pursuit-evasion differential games," *Robotics and Autonomous Systems*, vol. 59, pp. 22–33, 2011.

[7] K. H. Hsia and J. G. Hsieh, "A first approach to fuzzy differential game problem: guarding a territory," *Fuzzy Sets and Systems*, vol. 55, pp. 157–167, 1993.

[8] Y. S. Lee, K. H. Hsia, and J. G. Hsieh, "A strategy for a payoff-switching differential game based on fuzzy reasoning," *Fuzzy Sets and Systems*, vol. 130, no. 2, pp. 237–251, 2002.

[9] L. Jouffe, "Fuzzy inference system learning by reinforcement methods," *IEEE Transactions on Systems, Man, and Cybernetics Part C*, vol. 28, no. 3, pp. 338–355, 1998.

[10] K. M. Passino and S. Yurkovich, *Fuzzy Control*. Boston, Massachusetts: Addison-Wesley Longman Publishing Co., Inc., 1st ed., 1998.

[11] J.-S. R. Jang and C.-T. Sun, *Neuro-fuzzy and soft computing: a computational approach to learning and machine intelligence*. Upper Saddle River, New Jersey: Prentice-Hall, Inc., 1997.

[12] L. A. Zadeh, "Fuzzy sets," *Information and Control*, vol. 8, no. 3, pp. 338–353, 1965.

[13] B. Al Faiya, "Learning in Pursuit-Evasion Differential Games Using Reinforcement Fuzzy Learning," Master's thesis, Carleton University, Ottawa, ON, Canada, 2012.

[14] T. Takagi and M. Sugeno, "Fuzzy identification of systems and its applications to modelling and control," *IEEE Transactions on Systems, Man, and Cybernetics*, vol. SMC-15, pp. 116–132, 1985.

[15] L.-X. Wang, *A Course in Fuzzy Systems and Control*. Upper Saddle River, New Jersey, Prentice-Hall, Inc., 1997.

[16] J. Jantzen, Design of Fuzzy Controllers. Technical Univ. of Denmark: Technical Report (No:98-E864) Department of Automation, Hoboken, NJ, 1999.

[17] T. J. Ross, *Fuzzy Logic with Engineering Applications*. John Wiley & Sons, Ltd, 2010.

[18] X. Lu, "On Multi-Agent Reinforcement Learning in Games." Ph.D. Thesis Carleton University, Ottawa, ON, Canada, 2012.

[19] X. Dai, C. Li, and A. Rad, "An approach to tune fuzzy contorllers based on reinforcement learning for autonomous vehicle control," *IEEE Transactions on Intelligent Transportation Systems*, vol. 6, no. 3, pp. 285–293, 2005.

[20] L. X. Wang, *A Course in Fuzzy Systems and Control*. Englewood Cliffs, New Jersey: Prentice Hall, 1997.

[21] T. Takagi and M. Sugeno, "Fuzzy identification of systems and its application to modeling and control," *IEEE Transactions on Systems, Man, and Cybernetics*, vol. 15, pp. 116–132, 1985.

[22] J.-S. R. Jang and C.-T. Sun, *Neuro-Fuzzy and Soft Computing: A Computational Approach to Learning and Machine Intelligence*. Upper Saddle River, New Jersey: Prentice-Hall, Inc., 1997.

[23] R. S. Sutton and A. G. Barto, *Reinforcement Learning: An Introduction*. Cambridge, Massachusetts: The MIT Press, 1998.

[24] S. N. Givigi, H. M. Schwartz, and X. Lu, "An experimental adaptive fuzzy controller for differential games," in Proceedings IEEE Systems, Man and Cybernetics'09, (San Antonio, United States), Oct 2009.

[25] W. M. van Buijtenen, G. Schram, R. Babuska, and H. B. Verbruggen, "Adaptive fuzzy control of satellite attitude by reinforcement learning," *IEEE Transactions on Fuzzy Systems*, vol. 6, no. 2, pp. 185–194, 1998.

[26] R. Isaacs, *Differential Games*. New York: John Wiley & Sons, Inc., 1965.

[27] A. MERZ, "The homicidal chauffeur," *AIAA Journal*, vol. 12, pp. 259–260, 1974.

[28] T. Basar and G. J. Olsder, *Dynamic Noncooperative Game Theory*. New York: Academic Press, 2nd ed., SIAM Classics, 1999.

[29] S. H. Lim, T. Furukawa, G. Dissanayake, and H. F. D. Whyte, "A time-optimal control strategy for pursuit-evasion games problems," in Proceedings of the 2004 IEEE International Conference on Robotics and Automation, (New Orleans, LA), April 2004.

[30] S. Desouky and H. Schwartz, "Hybrid intelligent systems applied to the pursuit-evasion game," in Systems, Man and Cybernetics, 2009. SMC 2009. IEEE International Conference on, (San Antonio, TX, October 2009, pp. 2603–2608, 2009.

[31] S. Desouky and H. Schwartz, "Q(λ)-learning fuzzy logic controller for a multi-robot system," in IEEE International Conference on Systems, Man and Cybernetics, 2010. SMC 2010, (Istanbul, Turkey), October 2010.

[32] S. F. Desouky and H. M. Schwartz, "Genetic based fuzzy logic controller for a wall-following mobile robot," in American Control Conference, 2009 ACC 2009, (St. Louis, MO), June 2009, pp. 3555–3560, IEEE, 2009.

[33] M. Grześ and D. Kudenko, "Online learning of shaping rewards in reinforcement learning," *Neural Networks*, vol. 23, pp. 541–550, 2010.

[34] M. J. Matarić, "Reinforcement learning in the multi-robot domain," *Autonomous Robots*, vol. 4, pp. 73–83, 1997.

[35] J. Randløv and P. Alstrøm, "Learning to drive a bicycle using reinforcement learning and shaping," in 15th International Conference on Machine Learning, pp. 463–471, 1998.

Chapter 6
Swarm Intelligence and the Evolution of Personality Traits

6.1 Introduction

Swarm intelligence can be defined as the study of the social insect metaphor for solving problems in computer science or engineering [1]. It is a very powerful idea and a very interesting topic of research.

In this chapter, we study some of the ideas involved in the subject. We start by discussing the evolution of swarm intelligence and swarm-based robotics. We continue by discussing the necessity of the representation of the environment. Thereafter, we introduce the concept of personality traits as applied to swarm-based robotics.

6.2 The Evolution of Swarm Intelligence

Swarm intelligence grew out of the observation of social insect colonies. It emphasizes the distributedness and direct or indirect interaction among (relatively) simple agents [1]. The approach is also designed to be flexible and robust, because a large number of agents will make the whole system resistant to individual failures.

Multi-Agent Machine Learning: A Reinforcement Approach, First Edition. Howard M. Schwartz.
© 2014 John Wiley & Sons, Inc. Published 2014 by John Wiley & Sons, Inc.

The first and main application of swarm intelligence is in combinatorial optimization. It may be thought of as an alternative to classical approaches and, more specifically, genetic programming. Thus, it is also successfully applied to communication networks [1]. A more exciting approach is the trial of achieving artificial intelligence based on simple agents—a type of collective intelligence. In this sense, intelligence is perceived as the capacity of solve problems and not pure rationality.

One of the main ideas of swarm intelligence is the small amount of (formal) communication among the many individuals in the society. The concept is that the individuals would be able to get information directly from the environment just by observing the behavior of the other individuals. This was named *stigmergy* (from the Greek *stigma*: sting, and *ergon*: work) [2]. The implications of this idea are enormous. First of all, the need for the bandwidth of communication among the agents is largely reduced. Second, the representation of the environment turns out to be very important because it is through this representation (and not the world itself) that a robot communicates. And finally, when we walk toward a robotic representation, sensors become a key aspect of the implementation.

In robotics, the key advantage of swarm intelligence is its simplicity. Since hardware has advanced so much in the past decade or so, it has become possible to implement simple robots and test them in operation. Most important is the possibility to implement the algorithms in very simple chips, such as field-programmable gate arrays (FPGAs).

In this chapter, we deal mostly with the approach of swarm intelligence applied to robotics. We try to understand the nature of cooperation among them and try coordination algorithms that will provide a basis for future development.

6.3   Representation of the Environment

The way the environment is represented has a great influence on which techniques may be used to control the robots as well as in how the payoff tables used by game theory are implemented. In order to reduce the robots' computational requirements, the environment is represented by potential fields in a technique known as *co-fields* [3].

Co-fields are based upon cooperation through social potential fields [4]. Each robot is considered to be a particle with a given position in a fixed time. The environment, containing the other robots, obstacles, and enemies, is represented individually by potential fields. The potential fields represent attraction or repulsion among the objects. In the approach presented here, the meaning of the field is determined by the robot through the adaptation of personality traits as discussed in next section. For now, we may say that the algorithm the robot is running might prefer a downhill—where the robot looks for a low potential valley within its representation—or an uphill—where the robot pursues a peak—approach depending on the task it has and the personalities it developed.

For each robot, according the potential field created by any object it perceives, we may write the "force" [5] acting over it by

$$\vec{F}_T = \sum_{attractive} \vec{F}_i + \sum_{repulsive} \vec{F}_j \qquad (6.1)$$

where, again, the attractive and repulsive forces are defined differently for each robot. As a result of that, each individual robot perceives the world in a potentially very different way from the others, much like people. In Fig. 6-1, the actual configuration of the world and the ways each one of two robots perceives it are shown. As can be seen, robot A does not know anything about robot B, which, on the other hand, has a representation of its own state (position) that does not fit the actual data, and the other three objects are represented as a cluster (in the form of a big attractor) and not individually. The differences in the representations may result from the learning of new traits of personality or from noise in the sensors. Either way, the robot will have some outcome from its reading. Even if it means it will do nothing, this will come as a result of a computation. In Eq. (6.1), $\vec{F}_T$ is the resultant of the individual forces (attractive and repulsive) acting on the robot.

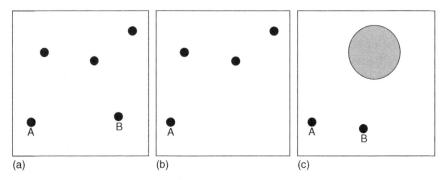

Fig. 6-1. (a) Actual configuration of the world. (b) The way robot A perceives it. (c) The way robot B perceives it. Reproduced from [21] © S. Givigi and H. M. Schwartz.

The direction of the robot's movement is the direction of the resultant force. Observe that the force may be represented as

$$\vec{F}_T = |F_T|\vec{d}_F \tag{6.2}$$

where $\vec{d}_F$ is the unit vector in the direction of the resultant force. In other words, it is

$$\vec{d}_F = \frac{\vec{F}_T}{|F_T|} \tag{6.3}$$

We use this notation in order to emphasize that the fields just define the direction of the movement, but the final decision to move in that direction and the velocity of the movement are determined by the personality traits as defined in the next section. The goal of the robot (equalize the sum of forces or get to a maximum/minimum value of them) is highly dependent on the task it has to perform.

6.4 Swarm-Based Robotics in Terms of Personalities

One of the suggestions of the theory of evolution is that animals have emotions [8]. Moreover, these emotions are shared by the same species. Also, *traits of personality* (a term that is used interchangeably with emotions) are important for the maintenance of objectives and collaboration [6]. Using these ideas, we can define a way robots may react to their environment [7, 9].

In our problem, the robots are assumed, initially, to be homogeneous in configuration and capabilities (understood as the set of traits of personality available to each one of them). However, like ants in a colony, they should differ from each other in order to better perform a complex task. But we do not want to add complexity to our algorithms. In order to solve this dilemma, we make use of personality traits. Therefore, the algorithms are the same for every robot, but changing these numerical values (the "traits") will change the behavior of individual robots. Although simple, this idea is very powerful and, when combined with reinforcement learning, may lead to a heterogeneous population in which some robots are able to specialize in executing certain tasks but at the same time they may learn how to execute a different action if it becomes necessary.

Personality traits are represented by real numbers γ_i. They are used in order to represent the individual intentions when faced with changes in the environment. The choice of actions is made considering the traits of personality a robot has and the payoffs related to each action at a given moment in time. Notice that the payoffs may change according to a change in the representation of the environment. We may give a human example to explain that. In our diet we usually would not consider eating worms. However, if we were lost in a jungle, we would do anything that would keep us alive. In the same way, a robot can change the values of its payoffs if the environment as it perceives it changes. Thus, the payoffs or rewards are dependent on the personalities of the robots.

Definition 6.1 *For a player i, we define the personality traits to be*

$$\bar{\gamma}_i = [\gamma_1, \dots , \gamma_n]^T \tag{6.4}$$

and the reward functions defined for each one of the n personality traits are represented by the vector

$$\bar{\mathcal{E}}_i = [\mathcal{E}_1, \dots , \mathcal{E}_n]^T \tag{6.5}$$

However, in this section we are going to drop the subscript i, since the algorithm to be introduced in the next section will be explained for just one robot.

The reward functions $\mathcal{E}_j$ represent how well a personality trait contributes to the success of the robot. These functions are arbitrary and defined based on the problem under consideration [9]. When an action α_k is chosen to be executed, all personality traits $\bar{\gamma}$ are then updated. The effect of the action taken is evaluated using the equation

$$\mathcal{V}(s, \bar{\gamma}, \alpha_k) = h \left(\sum_{j=1}^{n} \gamma_j \mathcal{E}_j(s_t, \alpha_k, t) \right) \tag{6.6}$$

where the individual reward functions $\mathcal{E}_j$ are related to how beneficial for the robot the execution of action α_k in the presence of state s_t is according to each personality trait γ_j and, therefore, determines the reward and/or penalty related to the trait of personality. Function $h(\cdot)$ is a suitably defined function that weights the cost function inside the summation in a way particular to each problem under consideration. Notice that the reward functions $\mathcal{E}_j$ are some feedbacks from the environment. Since there could potentially be more robots acting on the environment, the reward that one particular robot gets depends indirectly on the actions taken by the other agents present. Furthermore, we assume that the weights γ_j (i.e., the personality traits) are normalized, so

$$\sum_{j=1}^{n} \gamma_j = 1; \gamma_j \geq 0 \qquad (6.7)$$

However, notice that the personality traits vector $\bar{\gamma}$ is not a probability vector. In other words, γ_j is not the probability of the robot to take action j. Indeed, the dimension of the personality traits vector and the number of actions are, in general, not the same.

Equation (6.6) takes into account all the traits of personality γ_j and the environment as represented at time t (the current time step). Moreover, action α_k is the action under consideration and s_t is the state that the robot perceives the environment to be in at time t.

Furthermore, some procedure for evolving the personalities is necessary. The main purpose of that is to diminish the dilemma between stability and adaptability (plasticity) of learning.

The dynamics of the personality vector are described by the following general difference equation

$$\bar{\gamma}_{t+1} = \bar{\gamma}_t + \eta \bar{F}(\bar{\gamma}_t, \bar{\mathcal{E}}_t) \qquad (6.8)$$

where the function $\bar{F}(\bar{\gamma}_t, \bar{\mathcal{E}}_t)$ depends on the application under consideration.

Equation (6.8) implies, because the utility function is a function of all personality traits, that each trait of personality influences the others (notice that the term $\bar{F}(\bar{\gamma}_t, \bar{\mathcal{E}}_t)$ may include the reward functions for all the other personality traits). Therefore, changing a single trait will affect all the others. Furthermore, since the utilities include the actions of all other players (as explained above), the utilities and actions of other players also influence how the personality traits for one robot change.

The implementation of swarm-based robotics in terms of personalities is a very promising research topic. The implementation of such a technique is straightforward. It may also be shown that such a procedure results in coordination among a large number of robots [10].

In the following sections, some applications of the ideas presented so far are presented. Game theory and swarm intelligence are linked in the solution of some problems that may be related to multiple robotic environments, and algorithms and heuristics based on the idea of personality traits will be developed.

6.5   Evolution of Personality Traits

Swarm robotics has been defined by Sahin as *the study of how a large number of relatively simple physically embodied agents can be designed such that a desired collective behavior emerges from the local interactions among agents and between the agents and the environment* [11]. In our approach, we use a looser definition in what concerns the term *desired* collective behavior. We are more interested in what Beni [12] called *unpredictability* of the system. In other words, we are interested in the patterns and behaviors that will arise from the group interactions.

In practice, swarm intelligence is not intended to generate a rational individual entity as classical artificial intelligence proposes [1]. However, through investigation of simple computational models that may be implemented in simple machines, swarm intelligence tries to explore how complex tasks might be performed by a social entity because of behaviors that are not directly predicted by the particular characteristics of each individual [13]. For example, if we have a society with a majority of peaceful agents, when some aggressive individuals enter the community, we cannot say that the majority will make the new individuals become peaceful. It is our purpose to try to predict such behaviors through modeling and simulation in order to justify techniques that may be used in a swarm-based robotic environment.

In order to observe the emergence of group behaviors, we rely on the concept of "personality traits" [6, 7]. Using the adaptation of personality traits discussed, the underlying behavior of each robot changes, and by changing each robot's behavior the group's behavior also changes [1].

In order to demonstrate the ideas presented so far, we introduce three simulations that incrementally become more complex and together show how powerful is the approach suggested. We start, in Section 6.6, with a description of the general framework for simulation. In Section 6.7, we introduce a simulation of a zero-sum (matrix) game wherein no saddle point exists and therefore mixed strategies must be used. A theoretical proof of convergence is given, and the theoretical optimal solution is compared with the results obtained by the learning procedure presented in this chapter. In Section 6.8, we define some concepts that will be used in the solution of the problems presented in the following sections. Section 6.9 presents a more complex application than the one discussed in Section 6.7. In this section, we model a robotic conflict using a non-zero-sum game. Moreover, this time, we will have more players (three robots) and the reward will not only be the payoff table but a combination of payoffs and goal achievement. There is no proof of convergence for this case, and its existence is an open question. The results shown are based on heuristics. In Section 6.10, we introduce a situation in which the robots do not perceive the environment completely. Therefore, we will make use of potential fields for the representation of the environment and a still more complex learning procedure. This is also an open problem. However, the results are very promising. Finally, Section 6.11 concludes the chapter with a discussion of all simulations, bringing them together in a detailed manner.

6.6 Simulation Framework

All the simulations presented in this chapter will follow a single framework. However, the meaning of some of the terms will change as we go forward in considering more complicated situations. The general framework is described in Algorithm 6.1.

Algorithm 6.1 General algorithm

1: Define, if necessary, the variables necessary to represent the environment and initialize them. This may take several steps of the algorithm.
2: Initialize learning rate η. This value is dependent on the problem under consideration.
3: Define how many personality traits are necessary for the problem under consideration and initialize them.
4: Define the payoffs for the game. These payoffs may be a matrix (or set of matrices) as in Sections 6.7 and 6.9, or reward functions $\mathcal{E}_j$ representing the contribution of a trait of personality to the success of a mission combined with a matrix describing the state of the environment as in the simulation in Section 6.10.
5: Play the game. The rules for the game will be introduced in each simulation. In general, we use an equation in the form of (6.6).
6: Update the personality traits according to (6.28) and (6.29) in the case of the robots leaving the room and the robots tracking a target. In the case of the zero-sum game, the updating takes a slightly different form, which will be described shortly.
7: Normalize (if necessary) the personality traits γ_i for each robot according to (6.7).

For each of the simulations presented in this chapter, the steps in Algorithm 6.1 could be expanded as needed. Specially, in the simulations shown in

Section 6.10, each of the steps above is implemented as a series of several items in the algorithm.

6.7 A Zero-Sum Game Example

The game that we will analyze in this section is reported in Reference 14 and is represented in Table 6.1. This is a zero-sum game, meaning that the payoffs for each player always add to zero. For example, if player A plays the strategy A1 and player B plays the strategy B1, player A gets rewarded 4 units while player B gets a penalty of 4; whereas if player B decides to play strategy B2, then player B gets a reward of 4 units while player A gets penalized 4 units.

As may be seen from Table 6.1, there is no saddle point for the game, therefore there must be a mixed strategy set for both players. The optimal strategies, calculated using a linear programming solver such as the simplex method, for both players, are shown in Table 6.2.

6.7.1 Convergence

A general proof of convergence of strategies (or actions) to a Nash equilibrium is virtually impossible to be obtained for Eq. (6.8) of Section 6.4 or, equivalently, for the algorithm presented in Section 6.6. Therefore, we now provide

Table 6.1   Zero-sum game example.

		Player B strategies					
		B1	B2	B3	B4	B5	B6
Player A	A1	4	−4	3	2	−3	3
	A2	−1	−1	−2	0	0	4
strategies	A3	−1	2	1	−1	2	−3

Table 6.2   Optimal mixed strategies.

Player	Strategy	Optimal frequency (%)
Player A	A1	24
	A2	21
	A3	55
Player B	B1	0
	B2	36
	B3	0
	B4	57
	B5	0
	B6	7

a proof of convergence for the special case of zero-sum games.[a] Our proof will follow closely the one found in Reference 15. However, one fundamental difference exists. Since we use personality traits, the strategies' dynamics depend on them and additional considerations must be made. Therefore, personality dynamics are introduced in order to derive the strategies dynamics.

For a zero-sum game, the utility functions for the two players involved would be

$$\mathcal{U}_1(\bar{p}_1, \bar{p}_2) = \bar{p}_1^T M \bar{p}_2 \tag{6.9}$$

$$\mathcal{U}_2(\bar{p}_2, \bar{p}_1) = -\bar{p}_2^T M^T \bar{p}_1 \tag{6.10}$$

where $\bar{p}_1 \in \mathbb{R}^n$ and $\bar{p}_2 \in \mathbb{R}^m$ are arrays of probabilities where p_{in} is the probability that strategy n for player i be executed. The matrix $M \in \mathbb{R}^{n \times m}$ is the payoff matrix for both players. For the simulation at hand, matrix M is

$$M = \begin{bmatrix} 4 & -4 & 3 & 2 & -3 & 3 \\ -1 & -1 & -2 & 0 & 0 & 4 \\ -1 & 2 & 1 & -1 & 2 & -3 \end{bmatrix} \tag{6.11}$$

In order to proceed with our proof, we need to define some notations. This is done in the following definition.

Definition 6.2 *Consider $\bar{x} = [x_1, \dots, x_n]$, where n represents the number of strategies.*

- $\Delta(n)$ *denotes the simplex [16] in $\mathbb{R}^n$ (Fig. 6-2), that is*

$$\Delta(n) = \{\bar{x} \in \mathbb{R}^n : \bar{x}_i \geq 0 \; \forall i = 1, \dots, n; and \sum_{j=1}^n \bar{x}_i = 1\}$$

- *Int($\Delta(n)$) is the set of interior points of a simplex [16], that is*

$$int(\Delta(n)) = \{\bar{x} \in \Delta(n) : \bar{x}_i > 0 \; \forall i = 1, \dots, n\}$$

- *bd($\Delta(n)$) is the boundary of the simplex [16] (Fig. 6-2), that is*

$$bd(\Delta(n)) = \{\bar{x} \in \Delta(n) : \bar{x} \notin int(\Delta(n))\}$$

---

[a] In the particular case of zero-sum games, the convergence is to a Nash equilibrium.

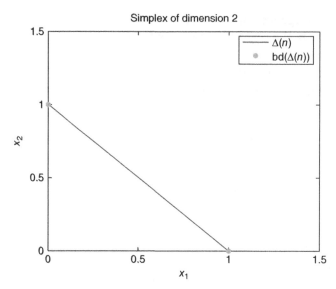

Fig. 6-2. Simplex of a player with two strategies. Reproduced from [21] © S. Givigi and H. M. Schwartz.

- $\bar{v}_i \in bd(\Delta(n))$ *is the ith vertex of the simplex* $\Delta(n)$, *that is*

$$\bar{v}_i = \{\bar{x} \in \Delta(n) : \bar{x}_i = 1 \text{ and } \bar{x}_j = 0 \ \forall j \neq i\}$$

Now we define the best response mappings as

$$\bar{\beta}_1(\bar{p}_2) = \arg\max_{\bar{p}_1 \in \Delta(n)} \mathcal{U}_1(\bar{p}_1, \bar{p}_2) \tag{6.12}$$

$$\bar{\beta}_2(\bar{p}_1) = \arg\max_{\bar{p}_2 \in \Delta(m)} \mathcal{U}_2(\bar{p}_2, \bar{p}_1) \tag{6.13}$$

The utilities in (6.9) and (6.10) are implementations of the utility function (6.6). We also need to define the reward functions related to each personality trait (the functions $\mathcal{E}_j$) and the function used to update the personality traits (function $\bar{F}(\bar{\gamma}, \bar{\mathcal{E}})$ used in (6.8)) for each of the players. Note that the arguments for the utility functions are apparently different from the parameters found in (6.6); however, the parameters are embedded in the definitions of $\bar{p}_1$ and $\bar{p}_2$, because they are dependent on $\bar{\gamma}_1$ and $\bar{\gamma}_2$, respectively (6.15). Function $h(\cdot)$ is the identity.

Let us now define the concept of empirical frequency (expectation of the opponent executing each of its actions). The empirical frequency $\bar{q}_i$ is calculated as the running average [15] of the observed actions of the opponent (recall that we have access to the action the opponent has played at each time step)

$$\bar{q}_1(k) = \bar{q}_1(k-1) + \frac{1}{k}(\bar{v}_{a_1(k-1)} - \bar{q}_1(k-1))$$

$$\bar{q}_2(k) = \bar{q}_2(k-1) + \frac{1}{k}(\bar{v}_{a_2(k-1)} - \bar{q}_2(k-1)) \tag{6.14}$$

where $a_i(k-1)$ is the action executed by player i at time step $k-1$ and $\bar{v}_i$ is a vertex of the simplex as defined in Definition 6.2. We may assume that as $k \to \infty$, $\bar{q}_i(k) \to \bar{p}_i$. Therefore, in the proof we are going to use both terms interchangeably.

Let us define a mapping from the personality space to the strategy space d : $\mathbb{R}^r \to \Delta(n)$ as

$$\bar{q}_1 = A_1 \bar{\gamma}_1$$

$$\bar{q}_2 = A_2 \bar{\gamma}_2 \tag{6.15}$$

Notice that A transforms from personalities to likelihood of actions. We can now define the reward functions for each personality trait, defined as (for each player)

$$\bar{\mathcal{E}}_1 = A_1{}^T (A_1 A_1{}^T)^{-1} \bar{\beta}_1(\bar{q}_2) \tag{6.16}$$

$$\bar{\mathcal{E}}_2 = A_2{}^T (A_2 A_2{}^T)^{-1} \bar{\beta}_2(\bar{q}_1) \tag{6.17}$$

where $A_1 \in \mathbb{R}^{n \times r_1}$ and $A_2 \in \mathbb{R}^{m \times r_2}$. These reward functions make use of the pseudo-inverse and are used in the convergence proof. In our example (with matrix M defined in (6.11)), $n = 3$ strategies and $m = 6$ strategies. Therefore, we select $r_1 = 5$ personality traits and $r_2 = 10$ personality traits. Notice that according to (6.16) and (6.17) $r_1 \geq n$ and $r_2 \geq m$. Moreover, $rank(A_1) = n$ and $rank(A_2) = m$. Other than those, no assumption is made. Furthermore, the functions used to update the personality traits ($\bar{F}(\bar{\gamma}, \bar{\mathcal{E}})$ in (6.8)) are

$$\bar{F}_1(\bar{\gamma}_1, \bar{\mathcal{E}}_1) = \bar{\mathcal{E}}_1 - \bar{\gamma}_1$$

$$\bar{F}_2(\bar{\gamma}_2, \bar{\mathcal{E}}_2) = \bar{\mathcal{E}}_2 - \bar{\gamma}_2 \tag{6.18}$$

The resulting personality dynamics are

$$\dot{\bar{\gamma}}_1 = A_1{}^T (A_1 A_1{}^T)^{-1} \bar{\beta}_1(\bar{q}_2) - \bar{\gamma}_1$$

$$\dot{\bar{\gamma}}_2 = A_2{}^T (A_2 A_2{}^T)^{-1} \bar{\beta}_2(\bar{q}_1) - \bar{\gamma}_2 \tag{6.19}$$

In this simulation, we relaxed the condition of normalization presented in (6.7). However, this is just for simplicity of calculations. Had we wanted to do so, this could have been easily implemented. Matrices A_1 and A_2 used were

$$A_1 = \begin{bmatrix} 0.3267 & 0.5071 & 0.7707 & 0.0478 & 0.3606 \\ 0.5406 & 0.7828 & 0.9703 & 0.1291 & 0.4767 \\ 0.1427 & 0.2456 & 0.3197 & 0.9082 & 0.2506 \end{bmatrix} \tag{6.20}$$

and

$$A_2 = \begin{bmatrix} 0.8686 & 0.6813 & 0.0693 & 0.2760 & 0.5695 & 0.5676 & 0.6390 \\ 0.6264 & 0.6658 & 0.8529 & 0.3685 & 0.1593 & 0.9805 & 0.6690 \\ 0.2412 & 0.1347 & 0.1803 & 0.0129 & 0.5944 & 0.7918 & 0.7721 \\ 0.9781 & 0.0225 & 0.0324 & 0.8892 & 0.3311 & 0.1526 & 0.3798 \\ 0.6405 & 0.2622 & 0.7339 & 0.8660 & 0.6586 & 0.8330 & 0.4416 \\ 0.2298 & 0.1165 & 0.5365 & 0.2542 & 0.8636 & 0.1919 & 0.4831 \end{bmatrix}$$

$$\begin{bmatrix} 0.6081 & 0.1034 & 0.1500 \\ 0.1760 & 0.1573 & 0.3844 \\ 0.0020 & 0.4075 & 0.3111 \\ 0.7902 & 0.4078 & 0.1685 \\ 0.5136 & 0.0527 & 0.8966 \\ 0.2132 & 0.9418 & 0.3227 \end{bmatrix} \tag{6.21}$$

The values of matrices in (6.20) and (6.21) were generated randomly. The point is that, if the matrices satisfy the conditions listed above, the algorithm will converge. More importantly, if A_1 and A_2 are the identity matrices of the necessary dimensions, the algorithm reduces to fictitious play and convergence is still guaranteed [17].

Using these definitions, we find that the strategy dynamics for player 1 is

$$\begin{aligned} \dot{\bar{q}}_1(t) &= A_1 \dot{\bar{\gamma}}_1(t) \\ &= A_1 (A_1{}^T (A_1 A_1{}^T)^{-1} \bar{\beta}_1(\bar{q}_2(t)) - \bar{\gamma}_1(t)) \\ &= \bar{\beta}_1(\bar{q}_2(t)) - \bar{q}_1(t) \end{aligned} \tag{6.22}$$

In the same way, the strategy dynamics for player 2 is

$$\dot{\bar{q}}_2(t) = \bar{\beta}_2(\bar{q}_1(t)) - \bar{q}_2(t) \tag{6.23}$$

We now define a function that measures the maximum possible reward for the players

$$V_1(\bar{q}_1, \bar{q}_2) = \max_{\bar{x} \in \Delta(n)} \mathcal{U}_1(\bar{x}, \bar{q}_2) - \mathcal{U}_1(\bar{q}_1, \bar{q}_2) \tag{6.24}$$

where $\max_{x \in \Delta(n)} \mathcal{U}_1(\bar{x}, \bar{q}_2)$ is the best "strategy"[b] that may be used by player 1. Using (6.12), we know that

$$\max_{x \in \Delta(n)} \mathcal{U}(\bar{x}, \bar{q}_2) = (\bar{\beta}_1(\bar{q}_2))^T M \bar{q}_2 \tag{6.25}$$

Therefore, from (6.24), using the definitions of utility functions in (6.9) and (6.10) together with the definition of best response mappings in (6.12) and (6.13), and collecting terms, one gets

$$V_1(\bar{q}_1, \bar{q}_2) = (\bar{\beta}_1(\bar{q}_2) - \bar{q}_1)^T M \bar{q}_2 \tag{6.26}$$

In the same way

$$V_2(\bar{q}_2, \bar{q}_1) = -(\bar{\beta}_2(\bar{q}_1) - \bar{q}_2)^T M^T \bar{q}_1 \tag{6.27}$$

Finally, we may say that $V_1(\bar{q}_1, \bar{q}_2) \geq 0$ and $V_2(\bar{q}_2, \bar{q}_1) \geq 0$ with equality if and only if $\bar{q}_1 = \bar{\beta}_1(\bar{q}_2)$ and $\bar{q}_2 = \bar{\beta}_2(\bar{q}_1)$.

Now we prove that the learning procedure will converge to the optimal solution. We start by the following lemma:

Lemma 6.1 Define $\tilde{V}_1(t) = V_1(\bar{q}_1(t), \bar{q}_2(t))$ and $\tilde{V}_2(t) = V_2(\bar{q}_2(t), \bar{q}_1(t))$. Then $\dot{\tilde{V}}_1(t) = -\tilde{V}_1(t) + \dot{\bar{q}}_1^T M \dot{\bar{q}}_2$ and $\dot{\tilde{V}}_2(t) = -\tilde{V}_2(t) - \dot{\bar{q}}_1^T M^T \dot{\bar{q}}_2$.

Proof: By definition (6.24)

$$\dot{\tilde{V}}_1(t) = \frac{d}{dt} [\max_{\bar{x} \in \Delta(n)} \mathcal{U}_1(\bar{x}, \bar{q}_2(t)) - \mathcal{U}_1(\bar{q}_1(t), \bar{q}_2(t))]$$

$$= \frac{d}{dt} [\max_{\bar{x} \in \Delta(n)} \mathcal{U}_1(\bar{x}, \bar{q}_2(t))] - \frac{d}{dt} [\bar{q}_1^T(t) M \bar{q}_2(t)]$$

$$= \nabla_{q_2} [\max_{\bar{x} \in \Delta(n)} \mathcal{U}_1(\bar{x}, \bar{q}_2(t))] \dot{\bar{q}}_2(t) - \dot{\bar{q}}_1^T(t) M \bar{q}_2(t) - \bar{q}_1^T(t) M \dot{\bar{q}}_2(t)$$

We now use the fact that [18] ([15], Lemma 3.2)

$$\nabla_{q_2} \max_{\bar{x} \in \Delta(n)} \mathcal{U}_1(\bar{x}, \bar{q}_2(t)) = \nabla_{q_2} \max_{\bar{x} \in \Delta(n)} [\bar{x}^T M \bar{q}_2(t)] = \bar{\beta}_1^T(\bar{q}_2(t)) M$$

[b]For a discussion on pure and mixed strategies, refer to [16].

which yields (using (6.22) and (6.26))

$$\dot{\tilde{V}}_1(t) = \bar{\beta}_1^T(\bar{q}_2(t))M\dot{\bar{q}}_2(t) - (\bar{\beta}_1(\bar{q}_2(t)) - \bar{q}_1(t))^T M\dot{\bar{q}}_2(t) - \bar{q}_1^T(t)M\dot{\bar{q}}_2(t)$$
$$= -(\bar{\beta}_1(\bar{q}_2(t)) - \bar{q}_1(t))^T M\bar{q}_2(t) + (\bar{\beta}_1(\bar{q}_2(t)) - \bar{q}_1(t))^T M\dot{\bar{q}}_2(t)$$
$$= -\tilde{V}_1(t) + \dot{\bar{q}}_1^T(t)M\dot{\bar{q}}_2(t)$$

Therefore, $\dot{\tilde{V}}_1(t) = -\tilde{V}_1(t) + \dot{\bar{q}}_1^T(t)M\dot{\bar{q}}_2(t)$. A similar derivation may be used for $\tilde{V}_2(t)$, yielding $\dot{\tilde{V}}_2(t) = -\tilde{V}_2(t) - \dot{\bar{q}}_2^T(t)M^T\dot{\bar{q}}_1(t)$. ∎

And, finally, we enunciate the convergence theorem:

Theorem 6.1   The solutions of the system of differential equations (6.22) and (6.23) satisfy $\lim_{t\to\infty}(\bar{q}_1(t) - \bar{\beta}_1(\bar{q}_2(t))) = 0$ and $\lim_{t\to\infty}(\bar{q}_2(t) - \bar{\beta}_2(\bar{q}_1(t))) = 0$.

Proof: We know from Lemma 6.1 that $\dot{\tilde{V}}_1(t) = -\tilde{V}_1(t) + \dot{\bar{q}}_1^T(t)M\dot{\bar{q}}_2(t)$ and $\dot{\tilde{V}}_1(t) = -\tilde{V}_2(t) - \dot{\bar{q}}_2^T(t)M^T\dot{\bar{q}}_1(t)$. Define the function $V_{12}(t) = \tilde{V}_1(t) + \tilde{V}_2(t)$. Taking its derivative, we have

$$\dot{V}_{12}(t) = \dot{\tilde{V}}_1(t) + \dot{\tilde{V}}_2(t)$$
$$= -\tilde{V}_1(t) - \tilde{V}_2(t)$$

Since $\tilde{V}_1(t) \geq 0$ and $\tilde{V}_2(t) \geq 0$ with equality only at the equilibrium point of (6.22) and (6.23), $V_{12}(t)$ is a Lyapunov function and the theorem follows the arguments. ∎

6.7.2 Simulation Results

We now present the results for the problem presented in the matrix in (6.11).

Suppose that both players are initialized with personality traits set to random values (in the interval [0, 1]). This means that both players start with a random probability of playing each strategy that is different from the optimal one shown in Table 6.2. The question we want to answer is: if we use personalities as described above, will the players learn to play the best mixed strategy? If not, will there be any improvement over time?

Algorithm 6.2 is followed by both players. As described in (6.20) and (6.21), we have created five personality traits for player 1 and 10 personality traits for player 2. Observe that this is not necessary and we could have used a much larger number of personalities (or, on the other hand, a smaller number greater than or equal to the number of actions) to characterize our player.

The utility functions are defined in (6.9) and (6.10). If we discretize the equation of dynamics for the personality traits ((6.22) and (6.23)), we end up with equations of the form (6.8) where, again, the value η may be called the *learning rate* and is set to a positive number, and functions $\bar{F}(\bar{\gamma}, \bar{\mathcal{E}})$ are defined as in (6.18).

Algorithm 6.2 Zero-sum game

1: $\eta \leftarrow 0.1$
2: $t \leftarrow 0.01$
3: Set the number of personality traits for player A equal to 5 and for player B equal to 10.
4: Initialize the personality traits $\bar{\gamma}_A$ and $\bar{\gamma}_B$ randomly.
5: Define the payoffs for the game to be according to the matrix in (6.11). Also, define the mappings from the personality traits spaces to the actions spaces according to the matrices in (6.20) and (6.21).
6: Player A initializes the empirical frequency of player B, p_2, as 0. Player B initializes the empirical frequency of player A, p_1, as 0.
7: **for** $i = 1$ to 5000 **do**
8:     Player A calculates the action to play according to its probability distribution $\bar{q}_1 = A_1\bar{\gamma}$. The action chosen is called a.
9:     Player B calculates the action to play according to its probability distribution $\bar{q}_2 = A_2\bar{\gamma}$. The action chosen is called b.
10:     Both players play their actions.
11:     The payoff for player A is M_{ab}, and the payoff for player B is $-M_{ab}$.
12:     Update personality traits as described in (6.8).
13:     Player A updates empirical frequency of player B according to (6.14).
14:     Player B updates empirical frequency of player A according to (6.14).
15:     Record the payoff for player B ($-M_{ab}$).
16: **end for**

We then ran the simulation described in Algorithm 6.2. The loop in line 7 was run for 5000 iterations. We then collected the final probabilities of using each one of the three actions at player A's disposal and each one of the six actions at player B's disposal as well as the value that player B obtained in each simulation (the value obtained by player A is just the negative of the one obtained by player B). The results are summarized in Table 6.3. One may notice that,

Table 6.3   **Experimental results obtained for both players.**

Player	Action	Optimal frequency (%)	Experimental frequency (%)
Player A	A1	24	22.04
	A2	21	23.93
	A3	55	54.03
Player B	B1	0	1.05
	B2	36	36.10
	B3	0	0.02
	B4	57	56.39
	B5	0	0
	B6	7	6.44

indeed, the procedure made the personalities converge to their optimal values. Furthermore, we observe that the disadvantageous actions B1 and B3 were almost never used. Moreover, action B5, which is dominated by action B2, was never used. The learning procedure even caught the subtlety with which B2 dominates B5. Furthermore, the average payoff for player B was 0.0710, which is very close to the theoretical value of the game, 0.07 [14]. Notice that for a zero-sum game the approach reduces to fictitious play. Therefore, convergence to the Nash equilibrium will always occur. This is not the case for the other simulations presented in this chapter.

6.8   Implementation for Next Sections

In the next two simulations, we are going to use the following implementation.

Going back to (6.8), we have to define how the function $\bar{F}(\bar{\gamma}, \bar{\mathcal{E}})_t$ will behave. Defining $\Delta\mathcal{E}_i(t) = \mathcal{E}_i(s_t, \alpha_j, t) - \mathcal{E}_i(s_{t-1}, \alpha_k, t-1)$, where α_j is the action taken at time t and α_k the action taken at time $t-1$, we define the step as

$$\Delta\gamma_i(t) = \frac{\Delta\mathcal{E}_i(t)}{\sum\limits_{j=1}^{n} \Delta\mathcal{E}_j(t)} \tag{6.28}$$

where n is the number of personality traits of a robot. Now we define the adaptation law as

$$\gamma_i(t) = \gamma_i(t-1) + \eta\Delta\gamma_i(t), \ 0 < \eta < 1 \tag{6.29}$$

Equations (6.28) and (6.29) imply, because of the presence of all personality traits in the denominator, that each trait of personality influences the others. Therefore, changing a singular trait will affect the way all the others work.

Furthermore, the convergence of (6.29) is highly dependent on the value of the learning rate η: if this is small, the convergence is too slow and the robots take too long to adapt to new situations; if it is too high, the system oscillates a lot around some value before it converges, and when it does, it has a large probability to go to a local maximum (minimum). In our simulations, we use a small value for this term so that we achieve a smooth convergence. Equation (6.29) is in the same form as used in the reinforcement learning literature. More details may be found in Reference 19.

When the state is recognized, the action is chosen according to the formula known as the *randomized strategy* [19], which is useful for leading the robot to explore new actions and not just exploit a learnt sequence of actions. For completeness sake, we will repeat the next equation, which demonstrates the randomized strategy:

$$P(\alpha_i|s) = \frac{k^{\mathcal{U}(s,\bar{\gamma},\alpha_i)/T}}{\displaystyle\sum_{j=1}^{m} k^{\mathcal{U}(s,\bar{\gamma},\alpha_j)/T}} \tag{6.30}$$

In (6.30), P is the probability that action α_i will be executed when the state is s in the presence of the personality traits $\bar{\gamma}$, where $\bar{\gamma} = [\gamma_1, \dots, \gamma_n]$ is a vector of personality traits (as introduced in Section 6.4). The term k is a coefficient that defines how often the robot explores new solutions or exploits the ones it already knows as better ones. When k increases, the probability that the robot explores new choices decreases, and vice versa. T is a temperature parameter inspired in the Boltzmann theory of statistical mechanics. It is desired that, over time, T decreases to diminish exploration [19]. The utility function $\mathcal{U}(.)$ is related to the action under consideration and the current state. $\mathcal{U}(.)$ is the long-term expected reward and not just the instantaneous one. This means that the decisions the robot takes are based on the expectation to solve the problem (find a target or perform a predefined task) and not just on the instantaneous reward. This difference is implemented in the personalities, because as robots learn over time, they will develop the capability to "predict" the payoffs of their actions. Moreover, n is the number of personality traits. And, finally, m is the number of possible actions that a robot may perform. In all simulations we will use $k = e$ (i.e., $\exp(1) = 2.7183$) and $T = 1$. $T = 1$ means that we do not reduce exploration as time goes by. $k = e$ means that the robots have a preference for exploitation of already learnt strategies but are also open to exploration [19].

6.9   Robots Leaving a Room

Our second simulation is similar to the one introduced and reported in Reference 9. The setup is as follows: Three robots of 1 unit in diameter are located in a room with dimensions 8×8 (corners at $(0, 0)$, $(0, 8)$, $(8, 8)$, and $(8, 0)$). There is a door centered at $(3, 8)$ and with dimensions so that just one robot may pass. Inside the room, there are three robots located at positions $(3 + 3\sqrt{2}, 8 - 3\sqrt{2})$, $(3, 6)$, and $(3 - 3\sqrt{2}, 8 - 3\sqrt{2})$, that is, at a distance 6 units from the door (Fig. 6-3). It is assumed that one robot knows the positions of the other two without any noise, and it is also assumed that the robots only move in a straight line from their initial position toward the center of the door with a fixed speed, 1 unit per second. The problem may be described as a game shown in Table 6.4, which has the payoffs for player A. The values X and Y in the table must follow the rule

$$X \in \mathbb{Z}^+ \text{ and } Y \in \mathbb{Z}^- \tag{6.31}$$

where the values of X and Y will depend on how the designer chooses to represent the environment. If the designer wants to enhance the action "Walk" for player A, then set $|X| > |Y|$. On the other hand, if the designer wants to enhance the action "Wait," then set $|Y| > |X|$. And finally, if both are considered to be at the same level, set $|X| = |Y|$.

Before we state the algorithm, we need to make some definitions.

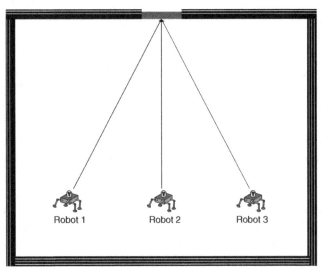

Fig. 6-3. Artistic depiction of the problem of robots leaving a room. Reproduced from [21] © S. Givigi and H. M. Schwartz.

Table 6.4   Modeling of a game between two robots trying to leave a room.

		Player B strategies	
		Walk	Wait
Player A	Walk	−1	X
strategies	Wait	Y	0

Definition 6.3   *Definitions for Algorithm 6.3.*

i. *The payoffs for the robots 1 and 3 in Fig. 6-3 are according to the matrix*

$$M_1 = \begin{bmatrix} -1 & 1 \\ -1 & 0 \end{bmatrix} \tag{6.32}$$

where $|X| = 1$ and $|Y| = 1$ (Table 6.4 and Eq. (6.31)). For player 2, the payoff table is

$$M_2 = \begin{bmatrix} -1 & 3 \\ -1 & 0 \end{bmatrix} \tag{6.33}$$

where $X = 3$ and $Y = -1$ (Table 6.4 and (6.31)). The values are different such that robot 2's expected reward is greater than its expected penalty.

ii. *The probability for a robot to move is given by*

$$P(Walk) = \frac{e^{\gamma_1}}{e^{\gamma_1} + e^{\gamma_2}} \tag{6.34}$$

where γ_1 is related to action 1, that is, "Walk," and γ_2 is related to action 2, that is, "Wait."

Algorithm 6.3 Robots leaving the room

1: Each robot will have two personality traits, initialized as $\gamma_i = \frac{1}{2}$, $i = 1, 2$, which define which strategy ("Walk" or "Wait" in Table 6.4) the robot will play.
2: Define the payoffs for the robots 1 and 3 according to (6.32) and the payoffs for robot 2 according to (6.33).
3: *Robots* ⇐ [1, 2, 3].
4: **while** there are robots in the room **do**
5:    For each robot calculate the probability to move according to the personality traits. The probability to move is given by (6.34), that is, $A_l \in$ {*Walk, Wait*} (where *l* is the robot's id).

6:   **for** $l \in Robots$ **do**
7:     **if** no other robot is the room **then**
8:       There is no conflict. Set action to "Walk," that is, $A_l \Leftarrow Walk$.
9:     **else**
10:      **if** $l = 1$ or $l = 3$, $M \Leftarrow M_1$ (6.32), otherwise $M \Leftarrow M_2$ (6.33)
11:      **for** $j \in Robots$, $j \neq l$ **do**
12:        $F(\gamma_{A_l}, \mathcal{E}_{A_l}) \Leftarrow M(A_l, A_j)$ {Payoff for robot l playing against robot j.}
13:        Update the personality trait related to the action chosen according to the equation $\gamma_{A_l}(t) \Leftarrow \gamma_{A_l}(t-1) + \eta F(\gamma_{A_l}, \mathcal{E}_{A_l})$.
14:      **end for**
15:      Normalize all personality traits γ_i so that $\sum_{k=1}^{2} \gamma_k = 1; \gamma_k \geq 0$.
16:    **end if**
17:    Add action A_l to list of actions L_l taken so far.
18:    **if** action is to walk and there is no collision **then** robot l moves **else** robot l keeps its current position for one time step.
19:    **if** robot l reached door **then** remove l from list $Robots$
20: **end for**
21: **end while**

The results were obtained after 100 repetitions of the game with a learning rate $\eta = 0.01$. First of all, one of the robots on the sides (robots 1 and 3) converged to a purely "cooperative" robot, that is, its personality trait for waiting for the others became 1; whereas the other robot on the side converged to a purely "competitive" robot, that is, its personality trait for always walking became 1. Second, the robot in the middle chooses its actions on a 50/50 basis. As result of all this, the average of the 100 games is 10.04 s to leave the room and the standard deviation is 1.82 s. Also, 24 out of the 100 repetitions obtained the best solution of 8 s.

One may notice that the robots did not work only for their own advantage. Table 6.4 shows that the strategy "Wait" is dominated by the strategy "Walk." However, behaving the way they did made the overall result much better for the group. This is one of the interesting results that will be exploited in the next simulation. Here we see the spontaneous emergence of altruistic behavior which enhances the performance of the group. The emergence of altruism is due to the calculations done in steps 13 and 15 of the algorithm. Notice that the matrices in (6.32) and (6.33) do not have a positive payoff for the strategy "Wait." However, since the traits that determine the execution of the actions are normalized (step 15), the negative payoffs that the strategy "Wait" gets combined with the negative payoffs of the collisions when strategy "Walk" is chosen will drive one of the robots to be altruistic.

6.10 Tracking a Target

We now make use of all the ideas presented so far and define a more complex and challenging simulation mission to be accomplished by several robots working together. We set up the simulation environment as shown in Fig. 6-4. In this figure, we depict a target (a tank) and several robots that are moving around it. Their objectives are to find the target and go back to the base. In our simulation environment, the position of the target changes from simulation to simulation and the robots perceive the environment as potential fields (Gaussian potential fields). Each single robot is able to identify the target potential field, the other robots' fields, and the base field. No noise is added to the readings, and some delay is possible in the measurements. We also assume that the low-level dynamics of the robots and the control loops necessary to stabilize them are already implemented.

Each robot has three traits of personality: "courage" (γ_1), "fear" (γ_2), and "cooperation" (γ_3), which influence which action the robot will take. For example, a courageous robot may pursue the gradient of the target, while a cooperative and fearful one may tend to huddle together with other robots in order to look for the target as a group. Again, these behaviors are derived from our assumptions on the definition of the "emotions" of the robots.

For this simulation, the environment is supposed to be in only two states: θ_1, meaning high risk for the robot (of being shot), and θ_2, which means that the robot is in a low risk of being shot. The decision about which state the robot is in is psychological, that is, it depends on the values of traits of personality

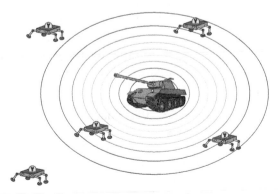

Fig. 6-4. Artistic depiction of the simulation environment. Reproduced from [21] © S. Givigi and H. M. Schwartz.

of each robot. In this way, if a robot is "courageous," high risk has a different meaning compared to a "fearful" robot.

Let $\sigma(.)$ be a function determining the threshold in separating states θ_1 and θ_2. Let also γ_1 be the trait "courage," γ_2 "fear," and γ_3 "cooperation." Define F_{Max} as the maximum potential field found so far. We then define the probability for the robot to identify the environment as state θ_1 (high risk) as

$$P(\theta_1 | s) = \frac{|F_T|}{|F_{Max}|} - \sigma(\gamma_1, \gamma_2, \gamma_3) \tag{6.35}$$

Since the traits of personality are normalized (as explained in the previous sections), we choose the threshold function to be

$$\sigma(\gamma_1, \gamma_2, \gamma_3) = \gamma_1 - \gamma_2 - \gamma_3 \tag{6.36}$$

Therefore, if the trait of personality γ_1 (courage) is dominant, the probability the robot will identify the environment as being "high risk" will decrease. On the other hand, since $P(\theta_2 | s) = 1 - P(\theta_1 | s)$, the probability increases when "fear" (γ_2) and "cooperation" (γ_3) are dominant. Notice that $P(.)$ could be out of the interval $[0, 1]$; if that happens, we simply truncate it.

In the same way, only two actions are possible. We will call them α_1, which means to follow an uphill approach (getting closer to the dangerous target), and α_2, which means to follow a downhill path (according to danger). Table 6.5 describes the payoffs related to each decision when the robot identifies the environment to be in each specific state. The values in Table 6.5 are empirical, and by choosing different payoffs the robots would end up with different behaviors. Also notice that the table is not exactly a payoff table as we had in the previous examples; in this case we do not have a conflict among the robots. The numbers in the table mean that, when the robot perceives the state to be in the "high risk" state θ_1, it is more "profitable" to execute action α_2 (downhill path), and when the robot finds itself in the low risk state θ_2, the robot would prefer to execute action α_1 (uphill path). Later on (in Eq. (6.37)), we will see

Table 6.5   Utility payoffs for states.

Utility payoffs		States	
		θ_1	θ_2
Actions	α_1	−1	5
	α_2	4	−2

that the choice is not so straightforward, but in general the rules just explained will be applied.

After the robot identifies the target, it gets back to the base with its estimation of the target location. The closer the robot gets to the target, the greater the danger of being shot (at each time step we divide the potential field where the robot is by the maximum value of the field-, the position of the target, and according to this number, randomly shoot at the robot simulating an action taken by the enemy). When the robot is shot, we assume that it is still operational, but has to go back to the base in order to avoid malfunctioning. Actually, since we may have a large number of robots, this assumption is not necessary, but by making use of it we simplify our simulation environment. When the robot is shot, we artificially increase its "fear" trait of personality in order to avoid being shot in the future. The task "get back to base" is hardwired in this approach, and after the robot identifies the target it just follows the track back to safety. Notice that this behavior is artificial and not desired, because the robot must be able to help other robots in need even if it is on its way back to the base. However, we do not implement this feature for the sake of simplicity.

The traits of personality are defined as follows:

1. Courage (γ_1): the robot goes in the direction of danger, that is, in the direction of the increasing potential field, therefore, this trait makes it more likely for the robot to identify the environment as in the "low risk" state (state θ_2 in Table 6.5).

2. Fear (γ_2): the robot goes in the opposite direction of danger, that is, in the direction of the decreasing potential field, therefore, this trait makes it more likely for the robot to identify the environment as in the "high risk" state (state θ_1).

3. Cooperation (γ_3): robots tend to huddle together in order to decrease the possibility of being shot. This trait makes the robots work together.

The behavior in (3) is explained by the assumption in the simulation that the chance of the robot being shot is inversely proportional to the number of robots huddled together. This is not a deliberate hypothesis; in fact, the same assumption has been made when studying the formation of patterns of animals in the wild (flock formation, fish schooling, etc.) [20].

To choose an action, we use the value function $\mathcal{U} : X \times A \to R$ in (6.37) which maps the state of the environment and the action under consideration to a reward. In the case of game theory, we need to calculate the expected value of the value function. Therefore, define $J(s_t, \bar{\gamma}, \alpha_i) = \sum_{j=1}^{3} \gamma_j \mathcal{E}_j(s_t, \alpha_i, t)$, that is, the summation in (6.6). Now, define $\mathcal{V}(s_t, \bar{\gamma}, \alpha_i) = \mathbb{E}\{J(s_t, \bar{\gamma}, \alpha_i)|(s, \alpha_i)\}$ as the expected value for the payoff for all possible actions α_i. We can think of this as a game against Nature [14], in which the environment is supposed to play with a mixed strategy $P(\theta_i|s)$. Therefore, the expected outcome of the game in Table 6.5 is

$$\mathcal{V}(s_t, \bar{\gamma}, \alpha_1) = \mathbb{E}\{J(s_t, \bar{\gamma}, \alpha_1)|(s_t, \alpha_1)\} = [-1(P(\theta_1|s) + 5(P(\theta_2|s)]J(s, \bar{\gamma}, \alpha_1)$$

$$\mathcal{V}(s_t, \bar{\gamma}, \alpha_2) = \mathbb{E}\{J(s_t, \bar{\gamma}, \alpha_2))|(s_t, \alpha_2)\} = [4(P(\theta_1|s) - 2(P(\theta_2|s)]J(s_t, \bar{\gamma}, \alpha_2) \quad (6.37)$$

Equation (6.37) is the application of game theory expectation calculation to the framework introduced in Section 6.4. Actually, this equation is just the implementation of (6.6) in terms of game theory, where $h(\cdot)$ is the expectation function.

Then, the action is chosen randomly based on the probability (6.30), where $k = e$ (exp $(1) = 2.7183$), $T = 1$:

$$P(\alpha_i) = \frac{e^{\mathcal{V}(s,\bar{\gamma},\alpha_i)}}{\sum_{j=1}^{2} e^{\mathcal{V}(s,\bar{\gamma},\alpha_j)}} \quad (6.38)$$

Before we state the algorithm, we need to introduce some definitions:

Definition 6.4 *Definitions for Algorithm 6.4.*

   i. *The target is identified by a Gaussian field. If (X_T, Y_T) designates the position of the target, let*

$$T(x, y) = Ke^{-\frac{1}{2}\frac{(x-X_T)^2}{\sigma^2}} e^{-\frac{1}{2}\frac{(y-Y_T)^2}{\sigma^2}} \quad (6.39)$$

   be the Gaussian field irradiated by it. σ is the standard deviation of the field and K is a term to scale the sensitivity of the robots.
   ii. *The robots also irradiate Gaussian fields. If (X_R, Y_R) is the position of a robot, let*

$$R(x, y) = \frac{1}{\sigma\sqrt{2\pi}} e^{-\frac{1}{2}\frac{(x-X_R)^2}{\sigma^2}} e^{-\frac{1}{2}\frac{(y-Y_R)^2}{\sigma^2}} \quad (6.40)$$

   be the Gaussian field around it. σ is the standard deviation of the field.

iii. *The uphill unit vector for robot i located at* (x_i, y_i) *is*

$$\vec{u}_i = \frac{\nabla T(x_i, y_i) + \sum_{j \neq i} \nabla R_j(x_i, y_i)}{|\nabla T(x_i, y_i) + \sum_{j \neq i} \nabla R_j(x_i, y_i)|}. \tag{6.41}$$

iv. *The downhill unit vector for robot i located at* (x_i, y_i) *is*

$$\vec{d}_i = -\vec{u}_i \tag{6.42}$$

v. *The probability for the robot identifying that it is in state* θ_1 *(high risk) is*

$$P(\theta_1|s_t) = \frac{|\nabla T(x_i, y_i) - \sum_{j \neq i} \nabla R_j(x_i, y_i)|}{|F_{Max}|} - \gamma_1 + \gamma_2 + \gamma_3 \tag{6.43}$$

where $|F_{Max}|$ *is the maximum field found so far for each robot. Accordingly, the probability to be in state* θ_2 *(low risk) is*

$$P(\theta_2|s_t) = 1 - P(\theta_1|s_t) \tag{6.44}$$

vi. *The probability of executing action* α_1 *is*

$$P(\alpha_1) = \frac{e^{\mathcal{U}(s,\bar{\gamma},\alpha_1)}}{e^{\mathcal{U}(s,\bar{\gamma},\alpha_1)} + e^{\mathcal{U}(s,\bar{\gamma},\alpha_2)}} \tag{6.45}$$

Accordingly, $P(\alpha_2) = 1 - P(\alpha_1)$. *Equation (6.45) is (6.30), where* $k = e$, $T = 1$, *and* $n = 2$.

vii. *The personality traits are updated using the adaptation law*

$$\gamma_i(t) = \gamma_i(t-1) + \eta \Delta \gamma_i(t) \tag{6.46}$$

where

$$\Delta \gamma_i(t) = \frac{\Delta \mathcal{E}_i(t)}{\sum_{j=1}^{3} \mathcal{E}_j(t)} \tag{6.47}$$

viii. *The probability of a robot being shot is*

$$P(shot) = \frac{|\nabla T(x_i, y_i) - \sum_{j \neq i} \nabla R_j(x_i, y_i)|}{\max(|T(x_j, y_j)|)} \cdot 0.01 \tag{6.48}$$

where (x_j, y_j) *are all the points visited by the robot previously.*

Algorithm 6.4 Tracking of a target

1: {Initializations} Define a base and set the initial position of all robots to it. Randomly select the position of the target (X_T, Y_T) and its standard deviation σ. Set $K = \dfrac{100}{\sigma\sqrt{2\pi}}$ and create the Gaussian field according to (6.39). Initialize each personality trait to a random value in $[0, 1]$. Normalize them so that $\sum_{j=1}^{3} \gamma_j = 1$; $\gamma_j \geq 0$.

2: For each robot located at the position (X_R, Y_R), define the field around it to be according to (6.40) with $\sigma = 4$.

3: Initialize a list $Robots \Leftarrow [1, 2, \ldots, n]$ with all robots.

4: **repeat**

5:    **for** $i \in Robots$ {for all robots} **do**

6:       Calculate the gradient $\nabla T(x_i, y_i)$ at the current position of the robot.

7:       **for** $j \in Robots$; $j \neq i$ **do** calculate the gradient of each robot's field $\nabla R_j(x_i, y_i)$.

8:       Calculate probabilities of identifying the robot in states θ_1 and θ_2 according to (6.43) and (6.44).

9:       Calculate the rewards for each personality trait $\mathcal{E}_1(.)$, $\mathcal{E}_2(.)$, and $\mathcal{E}_3(.)$.

10:      Calculate the expected values for the execution of each action according to (6.37).

11:      Calculate the uphill unit vector (6.41) and the downhill unit vector (6.42). Calculate the probability of executing action α_1 (uphill) and α_2 (downhill) as described in (6.45). Randomly select the action to be executed using these probabilities.

12:      Calculate the step for the adaptation of traits of personality according to (6.47).

13:      Update the personality traits using the adaptation law in (6.46).

14:      Calculate the probability of a robot being shot as in (6.48).

15:      **if** robot i is shot **then** remove robot i it from list $Robots$ and go back to base.

16:    **end for**

17: **until** all robots are back to base

As stated in Algorithm 6.4, the choice for the reward functions $\mathcal{E}_i(.)$ is dependent on the application. It may be argued that the reward functions would have to include some kind of external payoff based on the success of the task. However, this is not considered in the model of Algorithm 6.4.

The interpretation of the reward functions used in Algorithm 6.4 is as follows:

- $\mathcal{E}_1(.)$ is the function for the personality trait γ_1, "courage." For action α_1 "follow the uphill gradient," the reward is $\mathcal{E}_1(s_t, \alpha_1, t) = \nabla T(x_i, y_i) \cdot \vec{u}_i$. Notice that this value is positive if the angle between the gradient ∇T and the uphill unit vector $\vec{u}_i$ is in the interval $(-90°, 90°)$ and negative otherwise. For action α_2, "follow the downhill gradient," the reward is $\mathcal{E}_1(s_t, \alpha_2, t) = \nabla T(x_i, y_i) \cdot \vec{d}_i$. Notice that this value is positive if the angle between the gradient ∇T and the downhill unit vector $\vec{d}_i$ is in the interval $(-90°, 90°)$ and negative otherwise. In other words, the personality trait "courage" returns a larger value if the direction of the movement is closer to the gradient of the target. Since $\vec{u}_i$ and $\vec{d}_i$ are constituted by a summation of the gradient of the target and the gradients of the robots (Eqs. (6.41) and (6.42)), the direction that is closer to the danger is preferred.

- $\mathcal{E}_2(.)$ is the function for the personality trait γ_2, "fear." For action α_1 "follow the uphill gradient," the reward is $\mathcal{E}_2(s_t, \alpha_1, t) = (\sum_{j \neq i} \nabla R_j(x_i, y_i)) \cdot \vec{u}_i$. Notice that this value is positive if the angle between the summation of gradients $(\sum_{j \neq i} \nabla R_j(x_i, y_i))$ and the uphill unit vector $\vec{u}_i$ is in the interval $(-90°, 90°)$ and negative otherwise. For action α_2, "follow the downhill gradient," the reward is $\mathcal{E}_2(s_t, \alpha_2, t) = (\sum_{j \neq i} \nabla R_j(x_i, y_i)) \cdot \vec{d}_i$. Notice that this value is positive if the angle between the summation of gradients $(\sum_{j \neq i} \nabla R_j(x_i, y_i))$ and the downhill unit vector $\vec{d}_i$ is in the interval $(-90°, 90°)$ and negative otherwise. In other words, the personality trait "fear" returns a larger reward for the action that moves the robot closer to other robots.

- $\mathcal{E}_3(.)$ is the function for the personality trait γ_3, "cooperation." For both actions, the reward is calculated as $\mathcal{E}_2(s_t, \alpha_k, t) = \sum_{j \neq i} \nabla R_j(x_k, y_k)$, for $k = 1, 2$ where (x_k, y_k) is the future position of the robot. Let $\vec{p}_i$ be the robot's current position and $|\vec{v}_i|$ the speed of the robot, which in our case is 1 unit per second. For action α_1 "follow the uphill gradient," the reward function $\mathcal{E}_3(.)$ is evaluated at $(x_1, y_1) = (\vec{p}_i + \vec{u}_i \cdot |\vec{v}_i|)$. For action α_2, "follow the downhill gradient," the reward function $\mathcal{E}_3(.)$ is evaluated at $(x_2, y_2) = (\vec{p}_i + \vec{d}_i \cdot |\vec{v}_i|)$. The personality trait "cooperation" assumes that when the robot moves closer to other robots, the survival of the groups is enhanced.

Indeed, the concept of success in the definition of reward functions is very subjective. Depending on the information we have available and the complexity of the model we establish, success would have a completely different meaning. For example, with the same setup of Algorithm 6.4 we may assume that the

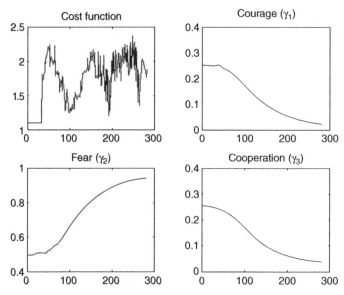

Fig. 6-5. Utility function and personality traits of one robot. Reproduced from [21] © S. Givigi and H. M. Schwartz.

robots know where the target is, and the task could be just to get close to it. In this case, the reward functions $\mathcal{E}_i(.)$ could include external information on how dangerous the environment becomes at each step, say how close a shot came to hit the robot, or how close we got to the target. Notice that in Algorithm 6.4 we did not assume that this information was available.

Since we chose a low value for the learning rate ($\eta = 0.01$), it is expected that there will be a slow convergence of the traits of personality to a steady-state value. Results for one arbitrarily chosen robot are depicted in Fig. 6-5. This figure indicates that the traits of personality do converge to a steady-state value. In this figure, the value function is $\mathcal{V}(s_t, \bar{\gamma}, \alpha_k)$, where α_k is the action executed at time t. Notice that the value function varies around some range (this is not necessarily the case; until further proof, this should be taken as a particularity of the simulation analyzed). We may notice that the robot becomes a "fearful" robot (γ_2 increases, while the other traits decrease). Therefore, we may hypothesize that this particular robot is in some kind of cluster of robots, which makes variations on the cost functions for the particular traits more difficult. Moreover, the particular values of the personalities are characteristic of the one simulation at hand. If we had a different initialization, we could get to different steady-state values for the traits of personality, since the environment changes considerably as well as the robots' initial conditions (the initial values for the traits of personality). Table 6.6 shows that in a given run all the robots

Table 6.6 Convergence of the personality traits.

Number of robots	Courage (γ_1)		Fear (γ_2)		Cooperation (γ_3)	
	Average	Standard deviation	Average	Standard deviation	Average	Standard deviation
10	0.1648	0.1377	0.1158	0.1435	0.7194	0.2743
20	0.2451	0.1006	0.2381	0.1713	0.5167	0.2316
30	0.1299	0.0930	0.3066	0.1827	0.5636	0.1692

Table 6.7 Simulation results.

Number of robots	Target location error		Total time		Location error for all robots	
	Average	Standard deviation	Average	Standard deviation	Average	Standard deviation
10	12.2000	6.5201	93.3000	55.6698	17.3810	7.7339
20	9.5880	3.8975	136.1000	45.4715	15.5337	5.4713
50	8.4136	3.9315	322.5000	117.8740	15.3012	5.5135

do converge to a steady-state value and they are related to each other. This is not necessarily true for different payoff tables (like Table 6.5) and reward functions ($\mathcal{E}_1, \mathcal{E}_2, \mathcal{E}_3$) and must be considered (until further proof) as a particularity of the simulation setup under analysis.

In order to evaluate the quality of the simulations, we measure the quality of the target location by the robots. When the ith robot goes back to base, it records the position (x_{S_i}, y_{S_i}) where it was shot (recall, we suppose that the robot just goes back to base when it is shot). Therefore, if (x_T, y_T) is the actual position of the target, the error of the best target location is $(\| (x_T, y_T) - (x_{S_i}, y_{S_i}) \|), i = 1, \ldots, n$, where n is the number of robots in the simulation. We also measure the total time it takes for all the robots to get back to base and the average location error for all robots in the simulation. The results are shown in Table 6.7, wherein we give the average and standard deviation of the target location error, total time of the missions, and average location error for all robots. All the results are obtained through 10 executions of the target-tracking mission.

The results indicate that the behaviors of some robots are independent of the number of robots in the fleet. There is also a tendency to get a better location of the target with increasing number of robots. This is due to our assumption that the robots are less likely to get shot when they are in larger numbers (Eq. (6.48)) because of huddling together, which has been observed in the simulations. In fact, in order to visualize better the effect of the other robots in how a robot decides to act, we considered the enemy (the tank) to be more accurate and replaced (6.48) by

$$P(shot) = \frac{|\nabla T(x_i, y_i) - 10 \sum_{j \neq i} \nabla R_j(x_i, y_i)|}{\max(|T(x_j, y_j)|)} \cdot 0.1 \qquad (6.49)$$

That is, the robots are 10 times more likely to be shot than predicted in the algorithm (therefore the probability is multiplied by 0.1 instead of 0.01). Also, the presence of other robots in the neighborhood makes it more unlikely for a robot to be shot (this is the meaning of the factor 10 in the equation above). In this way, robots will take advantage of the increase in the number of robots in the neighborhood. Table 6.7 also indicates that, as the number of robots increases, the total time for target location also increases, although not linearly. This happens for two different reasons: First, the robots take longer to leave the base (we assume that just one robot leaves the base at each time step). Second, as we have a larger number of robots, the chance of being shot decreases (again, (6.49)) and, therefore, they take much longer to get back to base.

When we use the probability in (6.48) (as in the simulation illustrated in Fig. 6-5), the traits "fear" and "cooperation" are always more important, giving rise to the most interesting behavior observed in the simulation, namely the tendency for the robots to huddle together. In most simulations, they formed a big group and kept as such until the individual robots were being shot by the enemy. This may also be seen in Table 6.7 because, as we increase the number of robots, the average distance to the target slightly decreases. This is also a result from the emergent huddling behavior. Figure 6-6 shows a picture of the state of the robots in the simulation. We can see that the robots do huddle together, but some of them (the more courageous ones) move farther from the center of the swarm. However, they are more likely to be shot (a result seen from (6.48)).

Fig. 6-6. State of the robots during the simulation. Reproduced from [21] © S. Givigi and H. M. Schwartz.

Another aspect observed in the simulations was the behavior of robots after some of them were shot. Observe that, since the number of active robots decreases, the reward $\mathcal{E}_2(.)$ for the personality trait γ_2, "fear," calculated in step 19 of Algorithm 6.4, and $\mathcal{E}_3(.)$ for the personality trait γ_3, "cooperation," calculated on step 20, decrease. Therefore, the reward $\mathcal{E}_1(.)$ for the trait of personality γ_1, "courage," gets more important for the remaining robots and they tend to "attack" the target more directly. This was a behavior observed when just some few robots were left. Since there is no other robot to help them, the remaining robot takes more risks and move toward the target, thereby increasing the risk of being shot.

In order to examine how resilient to spurious behaviors the swarm was, we fixed some robots as courageous and ran the simulation again. The idea was to check out when the group would start showing different group behaviors than the ones observed so far. Figures 6-7–6-9 are snapshots of the simulation over the period of time when some robots were set to "courageous." We also reduced the probability of getting shot even more, to just 10% of the value in

Fig. 6-7. State of the simulation when two robots turned courageous. Reproduced from [21]
© S. Givigi and H. M. Schwartz.

Fig. 6-8. State of the simulation when five robots turned courageous. Reproduced from [21]
© S. Givigi and H. M. Schwartz.

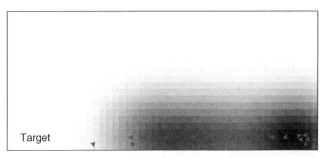

Fig. 6-9. State of the simulation when 10 robots turned courageous. Reproduced from [21]
© S. Givigi and H. M. Schwartz.

(6.49). Figure 6-7 shows the state of the simulation when 2 out of 20 robots were made courageous. It does not look very different from the state of the simulation in Fig. 6-6. Figure 6-8 shows the state of the simulation when 5 out of 20 robots were made courageous, and Fig. 6-9 shows the state of the simulation when 10 out of 20 robots turned courageous. We see that in Fig. 6-8 the group of robots start to break and, in Fig. 6-9, when half the robots turn courageous, the group of robots is completely broken. This suggests that the swarm of robots is resilient to outlier individuals up to some limit, but as the number of robots with some specific trait of personality increases, the swarm dynamics can dramatically change. In the cases depicted in Figs. 6-7–6-9, we see that, when more robots become courageous, they drive the entire group to a courageous state. We note that for this behavior to become noticeable we artificially and arbitrarily should set for the courageous robots the trait of personality γ_1, "courage," to 1 and the other two traits to zero.

The last interesting behavior that we want to discuss is how some robots turn around and follow other more courageous robots, that is, they wait until the courageous robots take the lead and then follow them. Figure 6-10 shows one robot that turned back and is waiting until a more courageous one passes by so it may follow the latter. The reason why this happens is that the traits of personality "fear" and "cooperation" are much bigger than the trait "courage." Therefore, the robot is (a) afraid of being shot and (b) wanting to share the risk with other robots.

6.11 Conclusion

This chapter has presented a unique method of modeling and controlling a swarm of robots. It integrates ideas from game theory and incorporates the

Fig. 6-10. Robot waiting for a more courageous robot. Reproduced from [21] © S. Givigi and H. M. Schwartz.

novel use of adaptive personality features to achieve an intelligent swarm. Three different simulations have been presented. Each simulation scenario highlights a different aspect of swarm intelligence using game theory and adaptive personalities.

The first simulation illustrated how two agents or robots can play a zero-sum game and how the agent/robot personalities would converge to the Nash equilibrium. A proof of convergence theoretically validated the method. The second simulation is an example of three robots that must cooperate in leaving a room. We showed how the proposed method could achieve optimal performance. Furthermore, one robot converges to the "always walk" condition, another converges to the altruistic "always wait" condition, and the third converges to the mixed 50% wait and 50% walk strategy.

The third simulation illustrated how the proposed method could be used to locate a target. The effect of different robot personalities on the performance of the swarm was shown. Cooperative robots tend to huddle into a tight swarm, whereas more courageous robots leave the swarm and lead the pack. We also demonstrated that the swarm was resilient to spurious individuals. By fixing some individuals as "aggressive," we showed that it takes up to half of the swarm to change the resulting group behavior. This is an important result because malfunctioning robots must be dealt with.

References

[1] E. Bonabeau, M. Dorigo, and G. Theraulaz, *Swarm Intelligence: From Natural to Artificial Systems*. New York, New York: Oxford University Press, 1999.

[2] P. P. Grassé, "La Reconstruction du nid et les Coordinations Inter-Individuelles chez Belli-cosistermes Natalensis et Cubetermes sp. La théorie de la Stigmergie: Essai d'interprétation du Comportement des Termites Constructeurs," *Insectes Sociaux*, vol. 6, pp. 41–80, 1959.

[3] M. Mamei, F. Zambonelli, and L. Leonardi, "Cofields: a physically inspired approach to motion cordination," *IEEE Pervasive Computing*, vol. 3, no. 2, pp. 52–61, 2004.

[4] R. Chalmers, D. Scheidt, T. Neighoff, S. Witwicki, and R. Bamberger, "Cooperating unmanned vehicles," in AIAA 1st Intelligent Systems Technical Conference, 2004.

[5] J. Borenstein and Y. Koren, "Real time obstable avoidance for fast mobile robots," *IEEE Transactions on Systems, Man, and Cybernetics*, vol. 19, no. 5, pp. 1179–1187, 1989.

[6] M. Mynsk, *The Society of Mind*. New York, New York: Simon & Schuster, 1986.

[7] S. N. Givigi and H. M. Schwartz, "A game theoretic approach to swarm robotics," *Applied Bionics and Biomechanics*, vol. 3, no. 3, pp. 131–142, 2006.

[8] C. Darwin, *The Expression of the Emotions in Man and Animals*. Chicago, Illinois: University of Chicago Press, 1965.

[9] D. Yingying, H. Yan, and J. Jing-ping, "Self-organizing multi-robot system based on personality evolution," in IEEE International Conference on Systems, Man, and Cybernetics, vol. 5, 2002.

[10] S. Givigi and H. M. Schwartz, "Evolutionary swarm intelligence applied to robotics," in Prooceedings of the IEEE International Conference on Mechatronics and Automation, pp. 1005–1010, 2005.

[11] E. Sahin, "Swarm robotics: from sources of inspiration to domains of application," in *Swarm Robotics: SAB 2004 International Workshop, Santa Monica, CA, USA, July 17, 2004: revised selected papers* (E. Sahin and W. M. Spears, eds.), (Berlin, Heidelberg), pp. 10–20, Springer-Verlag, 2005.

[12] G. Beni, "From swarm intelligence to swarm robotics," in *Swarm Robotics: SAB 2004 International workshop, Santa Monica, CA, USA, July 17, 2004: revised selected papers* (E. Sahin and W. M. Spears, eds.), (Berlin, Heidelberg), pp. 1–9, Springer-Verlag, 2005.

[13] M. Dorigo, V. Trianni, E. Şahin, R. Groß, T. Labella, G. Baldassarre, S. Nolfi, J. Deneubourg, F. Mondada, D. Floreano, and L. Gambardella, "Evolving self-organizing behaviours for a swarm-bot," *Autonomous Robots*, vol. 17, pp. 223–245, 2004.

[14] P. D. Straffin, *Game Theory and Strategy*. Washington, District of Columbia: The Mathematical Association of America, 1993.

[15] J. S. Shamma and G. Arslan, "Unified convergence proofs of continuous-time fictitious play," *IEEE Transactions on Automatic Control*, vol. 49, no. 7, pp. 1137–1142, 2004.

[16] J. W. Weibull, *Evolutionary Game Theory*. Cambridge, Massachusetts: MIT Press, 1995.

[17] J. Robinson, "An iterative method of solving a game," *Annals of Mathmatics*, vol. 54, no. 2, pp. 296–301, 1951.

[18] D. P. Bertsekas, *Dynamic Programming and Optimal Control*. Belmont, Massachusetts: Athena Scientific, 1995.

[19] L. P. Kaelbling, M. L. Littman, and A. W. Moore, "Reinforcement learning: a survey," *Journal of Artificial Intelligence Research*, vol. 4, pp. 237–285, 1996.

[20] R. Dawkins, *The Selfish Gene*. New York, New York: Oxford University Press, new ed., 1989.

[21] G. Sidney, "Analysis and Design of Swarm Based Robots Using Game Theory". Ph.D. Thesis, Ottawa, ON: Carleton University, Sept. 2009.

Index

Multi-Agent Machine Learning: A Reinforcement Approach, First Edition. Howard M. Schwartz.
© 2014 John Wiley & Sons, Inc. Published 2014 by John Wiley & Sons, Inc.

Printed and bound by CPI Group (UK) Ltd, Croydon, CR0 4YY

16/04/2025

14658364-0002